REVISITING ARMAGEDDON

Asteroids in the Gulf of Mexico

by

RAY COVILL

PUBLISHED
BY
BRIGHTON PUBLISHING LLC
435 N. HARRIS DRIVE
MESA, AZ 85203

Revisiting Armageddon
Asteroids in the Gulf of Mexico

by

Ray Covill

Published
by
Brighton Publishing LLC
435 N. Harris Drive
Mesa, AZ 85203
www.BrightonPublishing.com

ISBN 13: 978-1-62183-113-6
ISBN 10: 1-621-83113-2

Copyright © 2013

Printed in the United States of America

First Edition

Cover Design: Tom Rodriguez

No part of this publication may be reproduced or transmitted in any form or by any means, electronic or mechanical, including photocopy, recording, or any information storage retrieval system, without permission in writing from the copyright owner.

Revisiting Armageddon: Asteroids in the Gulf of Mexico

DEDICATION

*This book is dedicated to my family:
Ruth, Dave, Pete, and Lisa.
Without their help, this story would still be just a dream,
locked inside of my head.*

Ray Covill

TABLE OF CONTENTS

Acknowledgements
Foreword
Chapter One .... *Origins*
Chapter Two .... *Epic of Gilgamesh*
Chapter Three .... *Challenges to the Story*
Chapter Four .... *Clearing our Minds*
Chapter Five .... *The Gulf Event*
Chapter Six .... *Effects on North and Central America*
Chapter Seven .... *Plate Creep: Unscrewing an Answer*
Chapter Eight .... *Land of the Flat Rocks*
Chapter Nine .... *Creation of Magnetic North*
Chapter Ten .... *Florida Flattened and The Gulf Stream*
Chapter Eleven .... *A Wandering Delta*
Chapter Twelve .... *Yucatan and the Rio Grande*
Chapter Thirteen .... *Birth of the Bahamas*
Chapter Fourteen .... *The Asteroid Goes Underground*
Chapter Fifteen .... *Bubbles*
Chapter Sixteen .... *Rolling Limestone*
Chapter Seventeen .... *A Great Ship Goes Missing*
Chapter Eighteen .... *Transit Airlines Flight 981*

Revisiting Armageddon: Asteroids in the Gulf of Mexico

Chapter Nineteen .... *Why the Ship Sank*
Chapter Twenty .... *Time Shunts*
Chapter Twenty-One .... *The Formation of Florida*
Chapter Twenty-Two .... *Across the Pond*
Chapter Twenty-Three .... *Northern Flank of the Flood*
Chapter Twenty-Four .... *The Drowning of a Basin*
Chapter Twenty-Five .... *The Mediterranean Basin*
Chapter Twenty-Six .... *Southern Flank of the Flood*
Chapter Twenty-Seven .... *Wadi Kufra*
Chapter Twenty-Eight .... *Kinks in the Nile*
Chapter Twenty-Nine .... *Egypt's Great Sphinx*
Chapter Thirty .... *Basque Land*
Chapter Thirty-One .... *The Final Push*
Chapter Thirty-Two ....*Caves*
Chapter Thirty-Three ....*Noah*
Chapter Thirty-Four .... *Atlantis*
Chapter Thirty-Five .... *The Thief That Stole Mars*
Chapter Thirty-Six .... *Final Thoughts*
About the Author

ACKNOWLEDGEMENTS

A special thank you goes to Kathie McGuire and her staff at Brighton Publishing for placing their faith in a first-time author, and for editing my book and making suggestions and changes, all of which have hopefully made the story more appealing.

FOREWORD

I can find nothing in print that clearly explains most of the things in this story. I can only assume that I am the first to make an attempt at explaining the internal causal relationships of these events which took place so long ago. The things discussed in this book cannot, as I know them, be disproved; however, no one, including myself, can prove them with absolute certainty because there is no indisputable, factual information available to rely upon.

For these same reasons, I can't imagine any computer model capable of providing proof in either direction as computer results are no more reliable than the accuracy of the information entered into the computer. What's that old saying? "Garbage in, garbage out."

Nothing would thrill me more than to read about someone who can prove or disprove what I have written. I would eagerly thank anyone capable of refining my story—whether by reinforcing or negating what I have said. The more that others can add to the discussion initiated in my story, the more accurate it will become, no matter what direction it takes.

Certainly there are many people much more intelligent and more knowledgeable about sciences relating to this story, and I hope they will come forward to shed more light on this subject as time goes by.

But the basics of the story—the Gulf Event itself—seem factual to me, due to the many connections I have made.

This is a story which, for thousands of years, has been screaming to be told.

Read on.

CHAPTER ONE

Origins

There are several things about life on Earth that have puzzled me for years, but one in particular is the virtual absence of any information about life in Europe, North Africa, and the Middle East prior to the early Greek and Egyptian eras, meaning before 2500 to 3000 BC, or thereabouts. The Epic of Gilgamesh is the oldest written record, and the oldest oral history, the story of Atlantis, was eventually put into writing about 400 BC.

India, Tibet, and China have written records, usually preserved on stone or clay tablets, which trace back many thousands of years. To this day, Tibetan monks still work at deciphering, analyzing, and recopying the information on these tablets, in addition to their other work. No one seems to know when this translating and transcribing endeavor will ever be completed.

The Aborigine people of Northwestern Australia tell stories about the early times allegedly going back many tens of thousands of years. The Aborigines are descendants of the continent's earliest people and have lived in Northwestern Australia as long as the oriental people have done so in eastern Asia. As far as I know, the Aborigines have no written detailed, historical information; instead, they pass on oral stories and tales to each generation to explain their ancient beginnings and subsequent history.

Ten or eleven thousand years ago, some of the people of China, Mongolia, and Eastern Siberia suddenly left East Asia and crossed the Bering Straits into Alaska and then traveled south onto the North American continent. They did this in spite of extreme glacial conditions in North America, which coincidentally would have lowered sea level several hundred feet, making the Bering Sea crossing feasible.

What would have convinced these people to leave their traditional homelands and venture into the unknown? I can think of two possible explanations:

1. Famine resulting from war with neighboring people.
2. Some type of natural disaster of gargantuan proportion.

These races apparently thrived here in North America where they reproduced with one another and bred what we know as North American Indians. Some of these Native Americans, as we sometimes call them, are known to have maintained a written chronological history, pictorial in nature, of the major events in their lives from around eleven thousand years ago until about three-hundred years ago. At that time, the chronological history, recorded on a linen scroll of thirty feet or so in length, was in the custody of the Pokanoket Indians of southeastern Massachusetts and northeastern Rhode Island. The scroll was kept at the Pokanoket main camp in Bristol, RI.

King Philip's War, a bitter war between the English settlers and the Indians, ended in 1675 with an English victory and the death of King Philip, leader of the Pokanokets. Prior to the war, King Phillip was called Metacom, his Indian name.

Revisiting Armageddon: Asteroids in the Gulf of Mexico

During this war, the Native American population of southern New England suffered a loss of about 70%. Almost everything held by the Indians was destroyed by the white men, probably including the above described scroll since it hasn't been seen since the war's end. The chance that it still exists is remote, but it would certainly be fascinating to know what was on it.

The people of Central and Eastern Africa have their own ancient origin story, but I am not aware of anything in writing.

The first detailed records of a great Mediterranean civilization began to appear about 3,000 B.C. in Egypt in the form of pyramids and related stone carvings. Prior to this era, evidence of much older ancient times exists in the form of cave paintings in southern France, Egypt's Great Sphinx, Noah's Ark, and the Epic of Gilgamesh.

Chapter Two

The Epic of Gilgamesh

Gilgamesh, King of Uruk in Mesopotamia, came from an era forgotten until the 1800s when archaeologists began digging in 1839 to uncover some of the buried Middle Eastern cities.

This story goes back to the third millennium BC, or about 5,000 years ago. Gilgamesh is the first known tragic hero who stars in a story of adventure, reality, and tragedy. The *Epic of Gilgamesh* has been translated by many scholars of various languages over thousands of years. It is the first surviving epic poem, and the only one until Homer wrote the Iliad much later.

In the eleventh tablet, the *Epic of Gilgamesh* describes a flood, which some think may be the same flood described in the book of Genesis, but the entire relationship is controversial. There are similarities between the Genesis flood and the one told on the Gilgamesh tablets, but there are also striking differences. Whether they refer to the same flood event is conjectural. Several other ancient flood stories have survived which do not necessarily relate to the Genesis flood.

The *Epic of Gilgamesh*, in summary, tells of Gilgamesh and his family tearing down his house and building a boat. The

Revisiting Armageddon: Asteroids in the Gulf of Mexico

beam and length of the boat were about the same as those of Noah's Ark, at ten nindas, which would be 120 cubits. Similarly, the boat had seven decks. The top deck had a vaulted roof, and each deck was partitioned into nine compartments. The boat was allegedly built in seven days. Quite remarkable, if true. Once the ship was complete, Gilgamesh loaded his family and animals onto the structure.

Then came the rain and the storm. It would probably be referred to as a category five hurricane today.

It lasted six days. After seven days, Gilgamesh released a dove, followed later by a swallow. Both returned to the boat. He then released a raven, and it did not return. The boat eventually grounded on a mountain.

Although the story is similar to the biblical story of Noah, it is uncertain whether the two stories describe the same events. It could be two different floods, but there is no way to tell.

Chapter Three

Challenges to the Story

This story is based on factual information I have become aware of over the last few years. Not all the pieces of the puzzle are known to me, but some educated guesswork and the application of appropriate doses of common sense have helped fill in many of the gaps.

Assembling all of this information into a coherent document has not been an easy task because there were no precedents to guide me and no sources to turn to for information about the Gulf Event, which I am about to introduce. In addition, seemingly no one is even aware of the Gulf Event; no one—to my knowledge, that is—has put the topic into writing.

The scale of the Gulf Event natural disaster staggers my imagination, and more than once I considered abandoning the project altogether because the scope of its content is so unheard of that I felt no sane person would believe me. Yet, for 7,500 years, the story has gone untold and I feel it is a story the world should hear.

The Gulf Event drastically reconfigured the physical layout of parts of Central America, North America, North Africa, Europe, Western Asia, and destroyed at least one great civilization

Revisiting Armageddon: Asteroids in the Gulf of Mexico

(Atlantis).

When the flood waters abated, survivors from adjacent lands eventually repopulated these areas, which now appeared very different than before the flood.

Even though the Gulf Event provides answers to several questions and explains many mysteries man has pondered over for the last few thousand years, it has created even more questions than answers.

Chapter Four

Clearing our Minds

In many ways, we as humans are a product of our past. We are taught what our parents and educators have themselves learned and passed on to us. We hear about far away earthquakes, floods, fires, volcanoes, and other disasters, some of which are considered major in nature, or even "the worst in recorded history."

This automatically sets parameters in our minds as to just how severe any given type of disaster can be. These are extremes based on the detailed experiences of others, going back hundreds of years in some cases. But they are still only parameters. As a result of the things we know as truth, we often cannot imagine something a hundred or a thousand times worse. It is difficult to keep an open mind on the subject. I would like to think that I have been able to keep an open mind.

Nothing in the written history of our planet even partially explains what occurred on Earth 7,500 years ago. Students and scientists may be kept busy for a long time, delving into the ramifications of what you read here.

The Gulf Event actually happened—it is not a fictional adventure of my imagination. There are too many pieces of evidence that survive, which, taken as a whole, lead me to believe the event cannot be anything except factual, a piece of history, hitherto unknown.

The trick for me was to try and piece together a great many parts of the puzzle—sometimes tiny clues here and there that have a bearing on each other, and which, for the most part, are already known to historians, archeologists, and geologists. When these pieces can be shown to inter-relate with each other, some of the missing pieces start to fall into place and begin to make some kind of obvious sense. Soon the story began to develop into a believable connection of events.

We know that tsunamis have caused havoc, killed great numbers of people, and caused plenty of damage throughout the history of our planet. Most well-known or widely-reported tsunamis have occurred in the Pacific and Indian Oceans as a result of underwater earthquakes, or landslides in the seabed which jolt the surrounding sea and move great amounts of sea water rapidly away from the epicenter. When the resultant wave nears a shoreline, sometimes hundreds of miles away, the water slows due to the friction with the sea bottom in shallower depths. The wave then unloads itself onto the land and causes the detrimental effects we hear about in the media.

One of the largest disasters occurred in 1946, when a 55 foot tsunami wave struck the coast of Hawaii, killing many people and causing severe damage. On Dec. 26, 2004 a tsunami occurred off the west shore of Indonesia, causing floods on various islands in Indonesia, lands to the north, and all the way to the African horn near the mouth of the Red Sea, killing several thousand people, especially in the region east of India.

Floods occur regularly in some areas such as Bangladesh, causing great loss of life. The tsunami that struck this area in January of 2005 killed tens of thousands of people. Not only tsunamis, but cyclones (in the Atlantic Ocean we call them hurricanes), can also cause flooding and large death tolls.

In 797 AD an earthquake and flood caused great damage and a huge loss of life in Alexandria, Egypt, at the delta of the Nile River.

Major floods include the famous Johnstown, Pennsylvania flood in the late 1800s, numerous floods along the Mississippi river, and countless others around the world.

Perhaps one of the worst, in terms of human lives lost, occurred in China in 1931, killing four million people.

When I was very young, I lived in view of the New Bedford harbor and I witnessed the Great Hurricane of 1938 strike New England—or to be more precise, the New Bedford/Fairhaven, Massachusetts harbor and adjacent areas—without any warning, killing over 500 people and causing immense damage to boats, buildings, houses, docks, and roads. It was unlike any other hurricane we have ever seen in this area.

In more recent times, Hurricane Andrew caused billions of dollars in damage to Homestead, Florida, and in 2004 four hurricanes hit the Florida peninsula, causing 20 billion dollars in damage. The latest major disaster was Hurricane Katrina, which struck the New Orleans area, causing immense damage, killing a great number of people, and displacing thousands more. Several years later, things are still far from being resolved in New Orleans.

Numerous other hurricanes have devastated the United

Revisiting Armageddon: Asteroids in the Gulf of Mexico

States East Coast from Virginia to Texas on almost an annual basis.

A major hurricane hit Southeastern Massachusetts in the mid-1600s, but since the area was sparsely inhabited by white settlers, not much is known about it. We do know that the flood waters temporarily cut Cape Cod in half, and I believe this was the storm that permanently changed the Elizabeth Islands, west of Woods Hole, south of the Cape. Early maps, up to about 1610, show two major islands, but the storm changed them into four smaller ones, and they remain so to this day.

In addition to earthquakes, asteroids landing on or near large bodies of water can also cause tsunamis, sending waves of water crashing up to several hundred miles in every direction and affecting shorelines thousands of miles away. This is especially true if an asteroid comes in at an angle relatively high to the earth's surface, which could set the sea off in several directions. If the asteroid came in at around 90 degrees to the earth, the blast of debris and/or water would be uniform in every direction. Visualize a pebble dropping straight down into a puddle of calm water: the wave rings are circular and soon dissipate.

Great earthquakes much worse than the 1906 San Francisco quake have occurred, mostly along the "Pacific Rim," which includes the western shoreline of South, Central and North America, the Aleutians, Japan, China, the Philippines, and into Indonesia. Others occur in the Middle East, notably around the southern shores of the Black Sea, in Turkey and Iran. Sometimes the death toll is not great at the proximate locations of earthquakes, but when it comes to tsunamis, distant locations suffer from the effects, often resulting in loss of life.

Volcanic eruptions such as Vesuvius in 79 AD, have buried cities in Italy, Pompeii and Herculaneum, in hot ash. Few people escaped the ash-covered cities and most everyone died. Only in the last century has Pompeii been excavated, revealing a painful history through the ash-baked corpses of people in their last moments of life. The excavation also revealed a city frozen in time by ash, a rare detailed view of life 2,000 years ago in Italy. Mt. Pelee in the Caribbean Sea is another example of people and a town being buried in ash from a volcano. One person survived the disaster: a prisoner held in a cell deep underground.

More recently, Mt. St. Helens in Washington and other volcanoes in Central America and the Philippine Islands have left similar conditions.

Some volcanic eruptions spew out hot lava such as in Hawaii and Iceland, while others blast out ash, such as Vesuvius and Pelee. Others build up enormous pressure deep inside the volcano and then simply explode into the atmosphere. Santorini, also known as Thera or Thira, is an island in the southern Aegean Sea experienced this type of explosion in 797 AD in the Southwest Pacific Ocean, both in the 1800s. Krakatoa, or Krakatau, is a volcanic island situated in the Sunda Strait between Java and Sumatra in the Dutch East Indies. Tamboro is an active stratovolcano, also known as a composite volcano, on the island of Sumbawa, Indonesia. Because these eruptions were quite recent, more is known about them.

The explosion of Krakatoa was heard for thousands of miles. Many cubic miles of mountain were blown into the stratosphere to encircle the globe in the form of superheated volcanic ash. In New England, the ash dust was so dense that the sun could not warm the area, and it snowed in July for the only time in recorded history. That year was called "the year of no summer."

None of the natural disasters I have mentioned come close to wiping out an entire civilization and reshaping our land surfaces, but the Gulf Event did just that. It was the most recent geological disaster sustained by planet Earth, even though it occurred 7,500 years ago. When considering the existence of planet Earth in comparison to a human lifespan, 7,500 years ago for Earth is like five minutes ago to a person. The last natural calamity previous to the Gulf Event occurred about 100,000 years before, and nothing like it has happened since.

The rest of this book will delve into the details of what I call the "Gulf Event." I will begin in Central America, where it all began, and work easterly through the Atlantic Ocean, North Africa, the entire Mediterranean area, the Black Sea, and the Caspian Sea and beyond, where its easterly motion finally came to a halt.

Chapter Five

The Gulf Event

About 7,500 years ago, a string of major asteroids—at least six, and maybe as many as ten—most likely along with many smaller ones, had been drifting along together in the dark void of deep space for possibly many millions of years. It is also possible that they may have been orbiting the earth as part of a group of asteroids that still circle our planet today.

One may wonder why they didn't collide with each other and end up as one large asteroid, or drift away from each other onto separate courses. Asteroids are usually found to be composed in two ways: either loose concentrations of stony materials, or dense masses of solid rock with very high iron content. The latter would be much heavier than stony conglomerate material, much stronger, and resistant to fracturing.

As the asteroids traveled through space, a weak gravitational force resulting from their high iron content and consequent density tended to pull the asteroids together. Each asteroid would have had its own weak electrical charge from the time of its formation. They were kept from colliding with each other because they all held the same electrical charge, and like forces repel one another.

Revisiting Armageddon: Asteroids in the Gulf of Mexico

When two neighboring asteroids started to drift close to each other, they were pushed apart by their like forces, until gravity once again pulled them together. The push-pull forces would each have less and less effect on the asteroids until they eventually sorted themselves out and became more or less "locked" into a permanent position relative to each other. A few of these asteroids, for whatever reasons, began to fall out of orbit and approach Earth. If the Gulf asteroids had approached Earth only 50 to 100 miles higher, they might have missed us completely. The earth's geographical features and the locations and distribution of our inhabitants would have been drastically different compared to what we have today. We would also have several more species of animals, birds, and plants because many were undoubtedly lost forever in the blasts.

If the asteroid had missed us, and looking at the situation from an admittedly selfish point of view, I would not be here writing this story, and you wouldn't be here reading it, because there would be no story to tell. But such is not the case. Just the stuff of dreams, imagination, and "what ifs."

The order in which the asteroids landed is unknown, but it really isn't important. They came in low. Very low. About 5° above the Pacific horizon, heading ENE, or, to be more exact, 67° true. A few of them barely cleared the Sierra Madre range in western Mexico. It is even possible that they clipped off some summits and turned the debris into dust.

They began to land in the Central America region in fairly rapid succession. From here on, it may be helpful to refer frequently to the maps and diagrams in this book, or use your own detailed world map (I will use the names and places of land and water areas we all use today because, of course, no one knows the true original names of places the people of 7500 years ago may have used). Better yet, use a large globe to avoid the distortion a straight line on a flat map creates. A flat map does not account for the "Great Circle" effect one achieves through the use of a globe.

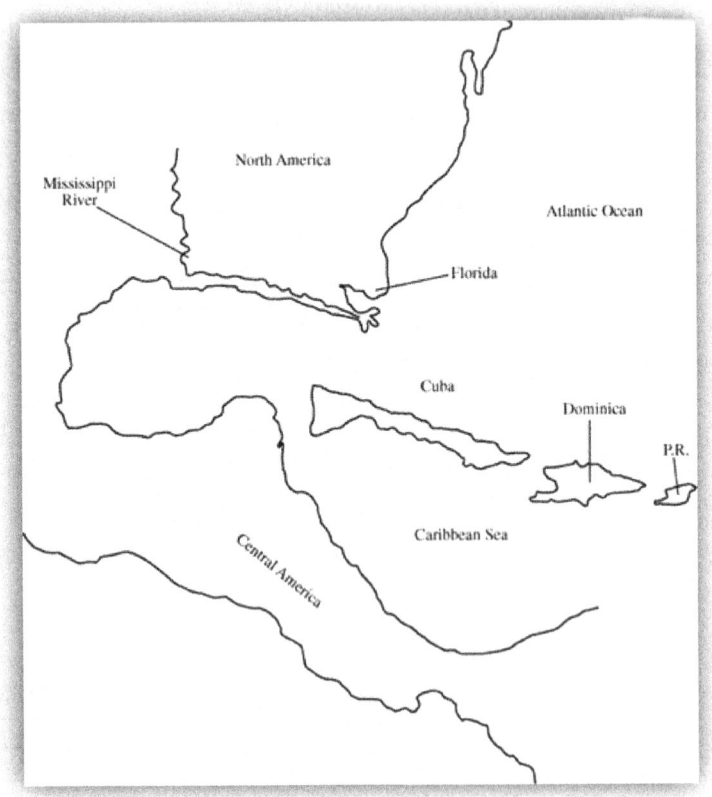

Gulf Region—Estimated Shoreline Prior to 7,500 YBP

A few of the asteroids were as small as five to twenty-five miles in diameter. It was one of these smaller asteroids that landed in the Gulf and created the Gulf of Campeche.

Another asteroid landed southeast of Yucatan, blasting out an elliptical area in the seabed called the Cayman Trench and creating the Gulf of Honduras. Its path ran ENE, parallel to the path of the huge Gulf asteroid. The Yucatan trench is about four hundred miles long and ends just south of central Cuba on the east side of Yucatan. Another asteroid struck further southeast off the north shore of Costa Rica, Nicaragua, and Panama, and opened up the Caribbean Sea down to the current east-facing coastline of southern Central America.

In the Pacific Ocean, one of the asteroids struck south of Panama and nearly severed the American continents from one another.

The largest asteroid landed in what is now the Gulf of Mexico, partially on land that existed at that time in the western parts of the gulf, and partially in the waters of the Caribbean Sea. It had the mass and speed to create an almost unstoppable momentum.

Based on what followed when it struck, I estimate it may have been roughly fifty or more miles in diameter. The asteroid had to be that large to create the changes to Earth's surface you will read about further in the story. The fact that it held together and did not break up on contact with Earth is one reason why I think it must have had a very high iron content. Also If it had been smaller, it wouldn't have had the necessary frontal area, weight, and momentum to force a 3,000 mile slice of the Atlantic Ocean to travel at incredible speeds from the landing point, the Gulf of Mexico, all the way to North Africa where the shallow water and sea bottom slowed the water and caused it to rise out of its bed. The overpowering wall of water charged into the Mediterranean Basin and filled it with sea water until the front of the surge slowed and finally stopped east of Armenia.

Ray Covill

The speed of these asteroids relative to Earth is unknown; Earth's gravitational force would have acted on these asteroids long before they entered our atmosphere and accelerated them to some unknown degree until impact occurred. However, based on the speed of Levy's Comet which struck Jupiter in 1994, and based on the effects on the Mediterranean area caused by the Gulf Asteroid, I estimate its speed to be 50,000 to 200,000 mph. To give you an idea of that speed, an asteroid moving at 200,000 mph relative to Earth, would travel the 3,000 mile distance from San Francisco to New York in just 54 seconds.

Chapter Six

Effects on North and Central America

Forces of contraction and expansion, especially along faults deep in the crust, build up over long periods of time. Even without the influence of asteroids, these tensions occasionally break loose in the form of earthquakes. Portions of the crust then shift slightly to release the tensions and shortly sort themselves out and settle down until the same thing occurs again at some time in the future. These effects, felt by people and machines on the surface above the faults, can be imperceptible or catastrophic, as we all know.

When a large asteroid crashes into our planet and tears it open, we suddenly realize that Mother Earth is really quite fragile after all.

The initial asteroid impact blasted the sea and dry land as well, throwing the earth's debris—rocks, soil, sand, water, and vaporized super-heated dust—in all directions, but due to the low trajectory of the asteroid's path, the majority of debris flew ENE. The blast created an elliptical hole we call the Gulf of Mexico, and caused earthquakes all over the planet.

Some of the debris could have easily been launched into space. Small particles of dust may be in Earth's orbit today in the form of belts around Earth, similar to the belts around Saturn, for example, only much less dense and less visible from Earth, or outer space.

At the speed the asteroid was moving, no animal or human could have seen it coming, even if they were looking in its direction at that moment. If a person were to look up as an asteroid passed low overhead, he or she might have seen a dark shadow cutting out the sunlight for a split second. And had the person not been looking up, the passage would have gone by unnoticed and silent. However, the case in point is very different because this particular asteroid was low enough to penetrate our atmosphere, and as it did so, a supersonic boom sent any loose surface materials in the immediate area adjacent to the blast sailing for hundreds of miles. Trees were torn from the ground, burned, or flattened instantly. People, plants, trees, birds, and animals in the path were destroyed instantly. The shock waves of the concussion reverberated and echoed around the planet many times before eventually subsiding.

The fact that the asteroids came in at a very low angle, and not 45° to 90°, plus the fact that they landed mostly in areas of sea water, all helped to minimize the resultant amount of dust in the atmosphere. Enormous amounts of "sea rain" fell for many weeks, bringing down a great deal of dust absorbed by the rain. To my knowledge, there are no geologic layers of silt found globally as evidence of a severe dust fall-out.

Some smoke and dust did encircle most of the planet, especially the Northern Hemisphere, for many months. As the dust blocked out the sunlight, temperatures began to fall, causing crops to fail. Consequently, food was scarce and animals died, or were killed off for food.

Revisiting Armageddon: Asteroids in the Gulf of Mexico

Bird migrations occur in most species all over the world; some for short distances and others for thousands of miles. I believe that migratory habits, including flyways, are firmly inbred into birds for generation after generation. However, it is possible that the Gulf Event created major disruptions in migration paths for birds flying through North America, Central America and South America. A certain experience comes to mind as evidence of this possibility. In late April a couple of years ago, while on a birding trip to the Dry Tortugas, seventy miles west of Key West, Florida, I found a very tiny, tired looking, Kentucky Warbler, about the size of a golf ball. I wondered how and why such a small bird, would attempt to cross almost a hundred miles of open ocean.

Prior to the Gulf Event, the Warbler's ancestors may have traveled that same distance, but the journey would have been mostly over land. They would have come up through South America to Yucatan, then up the peninsula (which was then a part of the Mexican mainland and which extended further north of its present northern terminus), to the entrance to the Gulf, (which would have been fairly small), and then to the Mississippi Delta, (which then extended easterly to a large area south of the Florida Panhandle). It would have been a safer route than today's path over the vast areas of ocean.

If this bird's present-day path of migration is from the Yucatan tip to eastern Cuba, continuing east to the Havana area, over seventy miles of ocean to the Dry Tortugas, and then to the tip of the Florida peninsula or up to the Florida Keys, the path would approximate its ancient route. The bird probably does not have the intelligence to follow the Gulf shoreline to the United States, and so the flyway over water is more probable. Ingrained habits dominate its behavior.

The American Indians, who lived in all parts of North America at that time, were decimated by the Gulf Event, and any survivors were forced to migrate south as the snow piled up and glaciers advanced in some places. The American Indians pushed south at first into the Southwestern and Southeastern United States, and then into Central America in search of a climate conducive to a better life. Their human necessities forced them to find places where the sun had broken through, allowing plant life to thrive once again. This glaciation period was the incentive for some of the natives to push further south into the South American continent.

The period of glaciation in Canada and the Northern United States continued for many years. The glaciers themselves built up to a thickness of one to two miles or more as they slowly crept downhill, following the natural contours of the land and deeply eroding the surface.

Eventually the skies began to clear, and the sun again warmed the land. Soon plant life resurged and animals and humans could reclaim their ancestral lost lands as the glaciers melted away. The glaciers left jumbles of rocks and gravel called terminal moraines at their frontal limits, and recessional moraines where materials in or on the glaciers dropped slowly in place as the glacier's forward motion ceased and the ice melted. These moraines are the evidence of their one time existence. The sun's increasing warmth plus heavy rains both contributed to the glaciers' demise.

Revisiting Armageddon: Asteroids in the Gulf of Mexico

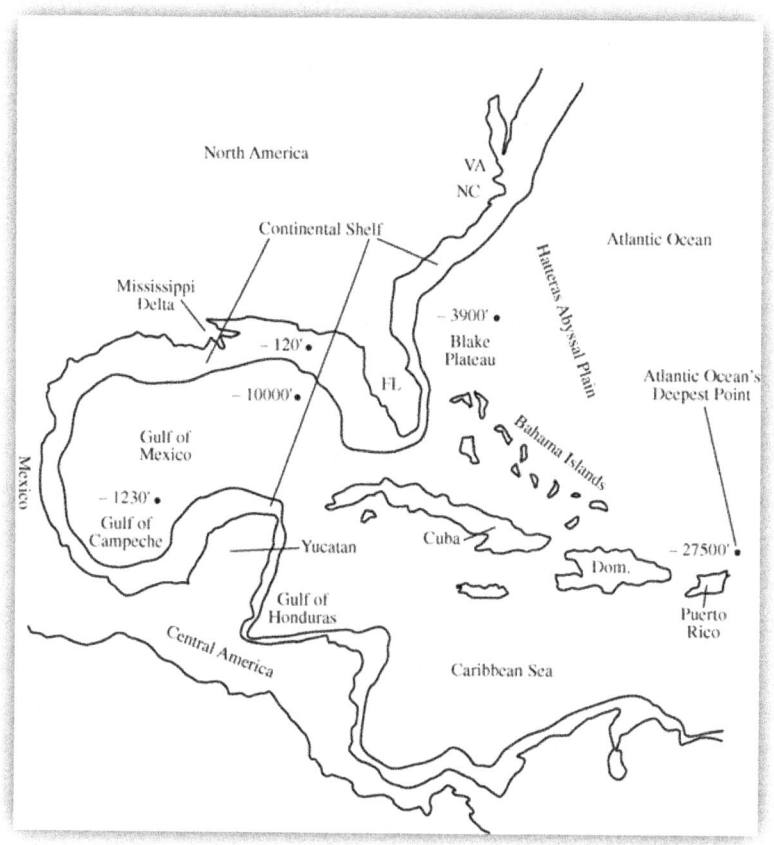

Gulf Region—After 7,500 YBP

Chapter Seven

Plate Creep: Unscrewing an Answer

When an asteroid the size of California or Alaska, or maybe even as large as the moon, strikes Earth in a weak place, especially on land as opposed to water, it could trigger a segment of Earth's crust, called a tectonic plate, to break loose from the edges of the crust and cause it to break into pieces. The plate then floats at a high rate of speed at first, over the molten lava or "magma" of Earth's interior. The plate sits on this magma until gravity and friction slow the momentum and settle it down again somewhere else. This shift would be very rapid at the start, and slowly decelerate until the movement could be properly called "plate creep."

In this process, tensions that have built up in the tectonic plates are relieved, and then subside to some degree until they build up again and at some time in the future another natural force springs them loose again. Plate creeps could last for thousands of years such as the former section of Antarctica that is now moving against Southeast Asia, or India, to be more specific. It is still pushing against and under the Himalayan Mountains, slowly raising their elevation. The momentum is still there, undoubtedly helped by Earth's centrifugal forces as the planet rotates.

Revisiting Armageddon: Asteroids in the Gulf of Mexico

In the early part of the twentieth century, scientists tried to determine the elevation of Mt. Everest and their calculations came out to exactly 29,000 feet above sea level. Because the number was too perfect, and because a number like 29,000 sounded too much like guesswork or rounding off, they announced the height at 29,002 feet. In recent years using satellites and more sophisticated technology, the height is now generally accepted as 29,030 feet. Everest is apparently still edging skyward as the Indian subcontinent continues to squeeze northward under the Himalayas.

However, Everest is not the tallest known mountain. That honor goes to a mountain that rises above the gentle rolling plane it sits upon. It is about 84,500 feet high, three times the height of Everest. It is called Olympus Mons, and it has never been climbed because it is located on Mars.

The tallest mountain on Earth, measured from summit to base, is not Everest either. Rather, it is Mauna Kea, the high point of Hawaii. At 13,850 feet, its base lies twenty miles out to sea and 4 miles down at the bottom of the Pacific Ocean, giving it a total gain of around 33,000 feet.

The remaining portion of Antarctica, covering the South Pole, is another example of tectonic plate shifting. Scientists have found within the rock and soil of Antarctica evidence of wormlike vegetation and organisms. Antarctica must have once been located some place further north in a more temperate climate for this to have happened, perhaps millions of years ago. I refer to this type of tectonic plate shifting as a "crust correction." To give you an idea of what happens, on a much smaller scale, place a large dried onion in your left hand, then place your right hand on top of it, and now unscrew it. It didn't work, right? But a very thin top layer of the onion should have quite suddenly broken loose in a number of places and shifted sideways until you stopped unscrewing the onion. This is similar to the crust correction phenomenon pertaining to Earth's crust as described above. The initial impact of the largest of the asteroids, referred to earlier, caused a unique effect on the Atlantic Ocean and the topography of the Central American region including the southeastern United States. The changes are, of course, permanent in nature, unless another asteroid disaster impacts the area at some time in the distant future.

The geographical changes in Earth's outer surface will never be reversed. Future asteroid impacts could only cause more changes and complicate the sorting out of the chronology of prior events similar to the Gulf Event.

There is no going back. Only forward, for better or worse.

CHAPTER EIGHT

Land of the Flat Rocks

The collisions of asteroids with Earth at very low angles, besides the Gulf Event, are not unheard of; in fact, they have occurred many times before all over Earth's land surface. Those that struck in the seas are nearly impossible to find and study. Other than the Gulf Event, one of the most obvious examples of an asteroid crashing at a low angle is in Canada's Hudson Bay.

The bay was created by an asteroid on a low trajectory coming in at 5° to 10° above the SSW horizon heading in a NNE direction. It neatly carved out a huge hole in the ground, blowing out a debris field to the NNE. Many islands in that area were thus created or enlarged. After the asteroid blasted the hole in Canada, small, flat pieces of this sedimentary rock landed west of the Greenland Straits on Baffin Island. Some of the rocks were large, flat rocks that covered a fairly sizeable area. I believe they came from the eastern side of Hudson Bay which was covered by a smooth, thin layer of sedimentary rock.

Ray Covill

Hudson Bay Rock Formation #1

The first explorers of European origin to see the flat rocks were the Viking explorers from southwest Greenland in around 995 AD. They marveled and wondered about the origin of these rocks, and aptly named the area "Helluland," meaning "land of the flat rocks."

A second, but smaller asteroid struck southeast of and adjacent to the Hudson Bay asteroid at about the same time. If you look at a world map, next to Hudson Bay you will find James Bay. Hudson Bay is larger than most of us realize. It is bigger than the Gulf of Mexico, and is almost the size of the eastern part of the United States, from the Mississippi River eastward to the Atlantic coast.

Revisiting Armageddon: Asteroids in the Gulf of Mexico

Hudson Bay Rock Formation #2

Off to the western shore of Hudson Bay, in the town of Churchill, Manitoba, one can find an unusual looking rock surface. This is either bedrock or molten rock that was dumped here during the asteroid impact. Smooth, flat rock can be seen all along the shoreline in this area. Ensuing glaciers have smoothed the surface so much that it almost looks as if it had been polished. There are many surface scratches, however, created over the last few billion years by glaciers grinding over in various directions. One could traverse parts of this landscape on roller skates, with a few jumps here and there over the larger cracks that have developed on the surface.

Chapter Nine

Creation of Magnetic North

Even though the Hudson Bay asteroid was moving to the ENE, there would have been some backlash of debris to the south, west and north. The force of this blast was even greater than that of the Gulf Asteroid. The Hudson Bay asteroid, made mostly of iron, dug a hole into the earth and buried itself into Earth's magma. This hole soon filled with sea water and debris, and exerted a huge magnetic force emanating from all the molten iron of which it was built. The bulk of it stayed in place near the earth's surface. There is no other evidence in that part of the Northern Hemisphere that gives us any clue that there could have been another contender for the Magnetic North Pole. The fact that it was so large caused its magnetic radiation to be felt all over the globe.

Maybe it is a good thing that this happened; otherwise today's ships and planes wouldn't be able to navigate so easily. Today's Global Positioning System devices can tell us our exact position on the globe. The only other system of navigation that comes to mind is the use of gyroscopes in navigation systems. In the ensuing years (probably several billion), the North American continent was caught up in a crust correction and slowly slid to the southeast. The bay, which was originally the entry point of the asteroid, came to rest—if indeed it has stopped sliding—and it is now hundreds of miles from where it was prior to the asteroid impact.

Chapter Ten

Florida Flattened and The Gulf Stream

Even though the Gulf asteroid was moving to the ENE, there must have been some backlash of debris to the south, west, and north. When the force of the blast dissipated, the sand, gravel, and stones soon landed near the trough created by the asteroid, ENE of the point of impact. By the time most of the debris fell back to Earth, the asteroid had torn a huge hole in the Gulf of Mexico, west of where the Florida peninsula now sits, and underneath the Atlantic a few hundred miles east of Florida.

The asteroid lodged deep inside Earth's crust and stopped either part way into the magma or completely into it. The intense heat from the magma melted the iron that was imbedded in the asteroid, and this caused certain activities in the ocean and atmosphere above, which are explained further along in the story. As it tore into Earth's crust, the asteroid was followed by a huge volume of sea water, possibly even a number of cubic miles. This sea water was trapped in Earth's crust and below in the form of super-heated water which was extremely high in temperature and obviously under enormous pressure.

Above ground, the falling debris spread out in a fan shape around the Tampa Bay region, quickly forming the peninsula. This is why the central part of the peninsula bulges out to the southwest toward the center of the impact area. The Florida peninsula, from the Okefenokee Swamp to Key West, was created by the Gulf Event.

After the asteroid passed through this area, the seas that were blasted to the north and south of the trough eventually slowed in momentum and came to a halt. Gravity then took over as the water which had been built up for hundreds of miles to the north and south came racing back to fill the void. When the waters from either direction met and collided, huge quantities of sea water burst in all directions—especially back into the Gulf of Mexico which ended up being quite flat and sandy due to the large amount of silt and sand being carried by the raging sea. As the seas calmed, the silt and sand settled over the Gulf. Today, very few large igneous rocks or boulders are found in the Gulf.

As the sea reversed and flowed over the Florida peninsula in every direction, the southern part of the peninsula was washed flat and left sloping down toward sea level until it disappears below the (new) sea level of the Florida Keys. This explains why the southern Florida landscape, including the Everglades, is so flat. In northern Florida there are gentle rolling hills and a few small rivers, but the flattening of the land is apparent the further south one goes down the peninsula.

Florida Bay is a continuation of this flattening procedure, and it becomes underwater a few feet at first, and more so as you move toward the Keys. The land on the west side of the peninsula gradually tapers off into the Gulf waters, which are quite shallow before suddenly dropping off to greater depths toward the south. To the east, the shallow lands east of the Florida Keys, made up of mostly fine material, besides coral and limestone, mark the southern limit of that portion of the debris field. I know of no other peninsula on Earth that is as large and as flat as Florida.

∽

The majority of the debris landed in Florida and off of Florida's east coast into the Atlantic Ocean. For the last 7,500 years the swiftly moving waters of the Gulf Stream have carried the material dumped off shore by the Gulf Event to the northwest, further up the coastline.

Ensuing major storms and hurricanes have created long barrier islands, close and parallel to the mainland. The water area between the two is referred to as part of the "inland waterway," a protected area where small boats can proceed safely along the United States mid-east coast all the way to Miami.

There are not as many barrier islands off the west coast of Florida, and they aren't as well-defined without the movement of the Gulf Stream. Additionally, they don't have as much exposure to the hurricanes coming off the Atlantic Ocean.

All the worlds' oceans have currents working along their shallow continental shelves. Since the Gulf is somewhat landlocked, no major ocean currents could develop in it.

My guess is that prior to the Gulf Event, the Gulf Stream came northwesterly along the north coast of South America, turned northerly between Yucatan and Cuba, and passed far east of the land and shallow inlets that were located in the Gulf of Mexico area, until it reached the Florida/Georgia shore. From there, it passed northerly off the East Coast until it turned northeast off Long Island, Block Island, Martha's Vineyard, Nantucket and Cape Cod. From here, shallow seas with sandy bottoms fan out to the northeast, as the now slower moving stream turns northeast

Revisiting Armageddon: Asteroids in the Gulf of Mexico

toward Europe.

After the sea had settled down once again, some weeks after the Gulf Event, the Gulf Stream would have entered the Gulf of Mexico. Here, the water entered a "dead end" area, because there was no outlet to the east coast of North America.

As the stream entered the Gulf, it poured its water into the Gulf and raised the sea level there higher and higher until it could push no more water into it.

The onrushing waters of the stream eventually found a course of lesser resistance in the areas north of Cuba, Dominica, and Puerto Rico to the Florida Keys. It was then jammed into a narrow gut between southeast Florida and the westernmost Bahamas. This funneling effect speeded up the stream's flow to the northwest until it slowed down along North Carolina and Virginia. As flowing water slows, it cannot carry mud, silt and sand as well as it did when it moved faster. This caused the Gulf Stream to dump a lot of suspended material off North Carolina and Virginia before it turned northward at a yet slower rate toward New England.

All this material created shallow "flats" off the coasts of North Carolina and Virginia. Subsequent major winter storms and hurricanes then easily built up barrier islands, trapping water and creating shallow areas to the west of the barrier islands, and thus formed Pamlico Sound and others off the United States East Coast.

This major shift in the course of the Gulf Stream was all due to the Gulf Event. The stream today follows that same course it took so long ago.

CHAPTER ELEVEN

A Wandering Delta

The current northern shoreline of the Gulf of Mexico has changed little since the Gulf Event, except for the Mississippi River Delta. The river existed at least 75,000 to 100,000 years, and probably even longer, before the Gulf Event occurred. Its delta extended further out into the Gulf of Mexico than it does today. The delta you see on a map today is "brand new," geologically speaking, having had only 7,500 years to form its present shape.

Because the river is still delta-building since the Gulf Event, it is quite accurate to assume it had been building a delta over that long expanse of time before the Gulf Event. But where? And in what direction? South to Yucatan? West to Mexico to Texas?

The answer lies in the current map of the delta. The meandering of the streams comprising the delta started out southerly, and is now bending slowly eastward. Look at a map and you will see what I mean. This is due in part to a lack of human intervention to dike it or deepen the channels.

The centrifugal force created by Earth's spinning on its axis, along with its direction of rotation, tends to throw the delta to

Revisiting Armageddon: Asteroids in the Gulf of Mexico

the east. For similar reasons, rockets sent up from Cape Canaveral to go into orbit, all head to the east to gain ground acceleration as Earth zips by below.

The flinging at the river's terminus moves to the east but we humans cannot detect it. Before the Gulf Event, the delta extended southeast and easterly to a point south of the Florida panhandle, and ended where it ran into the old Gulf Stream path well south of the current Georgia coastline. Remember, at this time we are referring to, Florida did not exist. This delta material was in the path of the northernmost part of the explosion.

The main portion of the blast, directly to the ENE of the center of the blast, would have dislodged earth from the bottom of the Gulf of Mexico which was then very differently shaped, and dumped the small-sized material in central and south Florida. These two general areas would also have been the most severely affected by the backwash of water rushing into the hole formed by the explosion.

The washing effect across this central and lower peninsula area continued for longer periods and in more directions because this area extends further southwest into the North Atlantic Ocean than the northern reaches of the peninsula. Florida's topography of higher elevations in the north and very flat lands in the south reflect the water's movement.

The Gulf Event explosion created an elliptical shaped hole with sharply-defined boundaries. It carved out a shoreline as clean as though it were cut out by an ice cream scoop, from Yucatan to Mexico to Texas, Louisiana, Mississippi, and the Florida panhandle. The Gulf has evolved into its present shape in the last 7,500 years and continues to grow and change each year as mud and silt from the river is deposited along the delta in times of flooding.

If you draw an imaginary line, east to west, across southern Louisiana where the Gulf of Mexico explosion appears to have reached its extreme northerly crater edge, the land south of that line amounts to about 8,800 square miles. All this new land has been built in the last 7,500 years.

It occurred to me that if I were to do some research in Louisiana records as to the average annual growth of the delta in square miles, I could determine whether that annual growth rate would have created the current delta over the last 7,500 years. Research and math wouldn't be sufficient because over the last hundred years or so, humans have interfered with the natural process of delta building by the erection of dikes to slow flooding, block off certain channels completely, and let most of the river descend through one large channel to the south terminus of the delta in order to create a deep water channel for the use of large commercial ships. However, I have shown on the accompanying map my guess as to the size and location of the Mississippi River's former course and delta prior to the Gulf Event. The delta probably mushroomed out a bit as its length grew. Look at a map of the Nile Delta in Egypt to see what I'm talking about.

The delta could have been the most prominent geological feature in the Gulf of Mexico area, along with Yucatan which was much wider and extended further to the north than it does today. The north shoreline of Cuba, west of Havana, extended more to the north and west than it does today so that Cuba and Yucatan were much closer at that time.

This delta, running easterly from Louisiana to its terminus

Revisiting Armageddon: Asteroids in the Gulf of Mexico

south of eastern Georgia, was made up of mostly mud and silt. When blasted by the asteroid's explosion, it wouldn't have traveled as far as the rocks. The silt came down south and east of Georgia and buried the northern end of the Florida peninsula.

It is interesting to note that the northern part of the Florida peninsula, starting at the Okefenokee Swamp, is today higher above sea level than the southern part, and contains rolling hills, lakes, and small rivers. As one goes further south, the land flattens out and approaches sea level. Southeast of the Everglades, the land goes underwater at Florida Bay until it terminates in the Florida Keys. I interpret this to mean that the asteroid dumped more debris in northern Florida, and less and less as the land slopes gently southward towards the Florida Keys.

Chapter Twelve

Yucatan and the Rio Grande

As a result of the Gulf event, the Yucatan peninsula of today is much smaller than it had previously been. The original Yucatan, prior to the asteroids' landing, was much larger, especially to the north where it almost met up with the then monstrous Mississippi River delta, and to the east where it neared Cuba. The exact configuration of the Gulf area is unknown, but a few clues do exist.

First of all, the Bay of Campeche was once part of Yucatan and southern Mexico—that much is fairly certain. The east side of Yucatan was struck by a different asteroid, which was probably the size of the one that created the Bay of Campeche. If it had been much smaller, land to the east would have survived. But none, of course, exists.

The northern shore of Yucatan is fairly straight, and runs to the ENE toward Cuba. This was the southern edge of the crater torn open by the Gulf Event. If you extend a line from this coast, it lines up almost exactly with the north shore of western Cuba (Cuba, of course, was created long before the Gulf Event). Apparently the Gulf Event's southern boundary not only defined Yucatan's north shore, but also redefined Cuba's north shore west of Havana.

～

The south side of the impact area affected two other areas of interest. First, the part of the Gulf that was south of the impact center, was a shallow bay that extended a few hundred miles westward where it ended in the small delta of the Rio Grande River. This would be a place well east of today's terminus near Brownsville, Texas. To the south, the Bay of Campeche was probably smaller than it is today.

The Rio Grande was formed long before the Gulf Event and was incised into its northern section by huge amounts of water draining the southern Rocky Mountain area. Originating in the mountainous regions, it carried much less silt than the Mississippi River, which drains mostly flat land of gravel and loam, and so the Rio Grande had less of a delta.

Chapter Thirteen

Birth of the Bahamas

The asteroid blasted away the now missing northern end of Yucatan and western end of Cuba. These lands were mostly made of stony materials, rather than pure sand. Due to the larger size of the debris, the impact of the asteroid caused this material to travel much further than sand or silt.

All this material landed haphazardly in an erratic hodgepodge east of the Florida peninsula and north of Cuba. This landing area was washed in every direction as the seas from the north and south collided, and eventually settled down, creating what we now know as the Bahamas. This makes the Bahamas the newest islands in the Caribbean Sea. Most of the other major islands are of volcanic origin, and were formed long before the Gulf Event.

There is one more piece of evidence in the Bahamas which backs up earlier statements about what happened when the Gulf asteroid imbedded itself in the earth and gases later found their way to the seabed along and above the crustal vent tunnels.

As the seabed was demolished and caved in by the asteroid, adjacent seabed areas were "stretched" upwards until they split apart in long straight lines parallel to the direction of the asteroid's path, and also in a few cases, perpendicular to the path. These cracks in Earth's crust weakened the surface region enough so that the molten magma, under terrific pressure, was able to rise up into these cracks, thus relieving some of the pressures deep below.

When the molten magma reached the seabed surface, it was immediately cooled by the ocean. Still, it was hot enough to flow into the cracks, filling them completely until the magma stopped erupting and then hardened by the cooling of the ocean. As it hardened, it contracted and shrunk, creating small elliptical and straight separations along the surface. The result was a long straight line of large sections resembling a path of "cobblestones," slightly resembling the roads in Italy built by the Romans, except that the "cobblestones" in the Bahamas were the size of Volkswagens.

If one could excavate along the sides of these Bahaman cobblestones, one would find no bottom to them. They would be all fused together in one long piece deep below the cobblestones.

When the Gulf asteroid landed, the vast majority of its frontal area hit sea water that existed there at that time, in addition to the Mississippi Delta mud, silt, sand and gravel. This material was sent into the atmosphere and then came down to create the Florida peninsula, and the eastern Gulf of Mexico. Excess material also landed in the offshore waters off Florida's east coast, much of which became suspended in the water of the Gulf Stream and was carried northwest to the United States mainland shores, as described earlier.

Chapter Fourteen

The Asteroid Goes Underground

The Gulf Asteroid struck in the central Gulf of Mexico area. Its low trajectory not only blasted sea water and earth high into the atmosphere and far out ahead of the impact point, but, more importantly, it was of such size and weight that it immediately buried itself in Earth's mantle. By the time it had traveled a few hundred miles, it had passed under the mantle at a point close to Florida's west coast. By the time it had moved beyond the east coast, probably several hundred miles off the east coast, it was completely through the mantle and into the interior molten magma.

The asteroid's high iron content contributed to its heavy mass and thereby minimized the extent to which it fractured on impact. It may have broken into several pieces, but was far from being obliterated into gravel, rocks, silt, sand, and dust.

The bulk of the asteroid tore a long elliptical hole in the Gulf area, blasting away sea water and earth. As it continued driving its way into Earth's crust for many miles, the sea, sent flying to the north, south and even to the west to a lesser degree, came quickly roaring back into the hole the asteroid created. The asteroid most likely didn't make it any further than partway into the magma.

Above the seabed, the seas churned around, in all directions, carrying along a great deal of sand, mud, gravel, and stony material which it deposited in the hole above the water that had fallen in behind the asteroid.

In addition, gravity collapsed the sides of the open hole. The hole became clogged up with all this sand and mud, effectively filling up the hole. The pressure of the sea water above sealed the hole. The sea water trapped below the debris in the hole and on top of the asteroid became subject to instant enormous pressure and temperatures; so great, in fact, that there was no space for the water to expand into steam, and it thus remained in its original liquid form. It became impregnated with all the typical characteristics of magma, meaning that it contained sulfur and other noxious elements which arose from the now molten iron asteroid, lodged forever within and under Earth's crust.

THE ATLANTIC SLICE

The most important part of the Gulf Even is what happened to the sea water east of the impact. Water, of course, is fluid in nature, and movement of the molecules has a domino effect or ripple effect. If the asteroid struck on dry land, the result would be a large explosion, but in the water, the force of the crash traveled from one water molecule to the next, driving the sea forward for hundreds of miles.

By the time the asteroid finally buried itself in Earth's magma, an entire slice of the North Atlantic Ocean was already several miles away, traveling ENE at incredible speed. The sea received such a gigantic and continuous push, that this slice of water—which could have been up to 100 miles long from north to south—eventually gained and maintained its own momentum and continued on until it struck land in Morocco, North Africa. The water wall may have been dozens of miles high as it approached the islands off the northwest shore of Africa and continued into the Mediterranean Basin.

One may be tempted to compare this high-speed traveling water wall to a tsunami, but to put it into perspective, consider the size of a typical tsunami, which is one or more large waves, moving away from a center point. The wave itself may be only a few hundred yards or so from front to rear. As it crashes onto land, the wave dissipates quite rapidly because nothing is giving it a constant push forcing it any further.

In the case of the Gulf asteroid, what started out as a slice of sea water around a hundred miles in width, slowly and exponentially expanded, until, by the time it hit Morocco, its main force may have been 300 to 500 miles wide with water north and south of the central portion extending several hundred miles on both sides.

This slice of ocean set in motion toward the Mediterranean area was 3,000 miles from front to rear, the distance from the Central Gulf to Morocco.

CRUSTAL VENT TUNNELS

Going back to the molten iron and superheated water mentioned earlier, it is important to think a bit about how a live steam reacts under pressure. A pressure vessel, such as a commercial heating boiler, contains a safety release valve and usually a low water cutoff. If the pressure of the steam begins to approach the maximum pressure the vessel was designed to safely contain, the safety valve is released and the steam is safely vented off to protect the boiler. If the safety valve should fail, the pressure could build until the entire vessel explodes which could destroy the building which houses it, kill or injure workers in the vicinity, and is audible for miles around.

A disaster of just this type occurred when a large manufacturing plant boiler exploded in New Bedford, Massachusetts, in 1954. The explosion destroyed the building housing it, killed five men, and was heard in Westport, Massachusetts, fifteen miles away.

The water trapped in Earth's crust in the aftermath of the Gulf Event explosion would have reacted in a similar manner, but on a much larger scale. As pressure built up, the water in certain areas would have sought out the paths of least resistance, rising to the surface of the seabed, and meandering until it managed to break through a hole (more likely hundreds of holes) in the seabed and blow out excess pressure as steam. I call these holes "crustal vent tunnels." After the water had relieved itself, it would subside and the pressure of the surface sea water would then jam the hole openings tightly shut, leaving no trace as to what happened.

In the early years following the Gulf Event, the crust vent tunnels off Florida's east coast would have peppered the ocean bottom and literally boiled the sea water for many years.

New occurrences of water escaping could, in some instances, reopen the original holes, or if water had built up elsewhere after shifting around deep in Earth's crust, new holes could have opened, creating escape tunnels where none had existed before. They can reoccur many times in one area if the crust is weak or permeated with prior tunnels, which would make that area a place of least resistance for steam releases to find and follow.

There must be an enormous volume of trapped sea water under the Florida offshore seabed in order to have caused frequent eruptions for 7,500 years. Many cubic miles of super-heated water must have been trapped below the surface to account for the frequent eruptions still taking place today. The fact that such action continues today is a remarkable indication of the strength and frequency of past eruptions.

As the centuries passed, the frequency of eruptions rate would most likely have decreased along with the decreasing volume of trapped water. Still, no one really knows if the frequency of eruptions is actually decreasing each year. One year or even one century is an insignificant length of time to consider these eruptions when they have been happening for 7,500 years. There is no reason to believe this activity will subside any time soon; it may go on for another 7,500 years, or possibly even longer.

But some day in the future, it must and will end. Just when that will take place is anyone's guess. I, for one, cannot estimate an end because there is absolutely no available factual information to provide any clues to make such a prediction.

One may be wondering about the power of these eruptions, and it is safe to say they vary greatly in location east of Florida and north of the Bahamas, the diameters of the areas affected, length of time they erupt, the concentration of crustal vent tunnels in any given eruption, and the frequency of eruptions.

It is entirely possible that some eruptions are tiny in size and affect only small areas. These would be far too small, especially in deep water, to affect any plane or boat above. Occasionally, however, several large eruptions, possibly

Revisiting Armageddon: Asteroids in the Gulf of Mexico

thousands, occur simultaneously in a concentrated area. If a boat or plane is in an area where such conditions exist, this raises the likelihood that a crash, sinking, or disappearance could occur.

If there were no boats or planes in the area, the conditions would abate and no one would ever know that the hazard existed. I think it is logical to assume that there must be many such occurrences, because at any one time there are very few boats and planes in the Bermuda Triangle area in any one place. The ratio of boats and planes per square mile must be tiny. The ocean is a very big place.

Much of the literature written since World War II, dealing with the disappearance of ships and planes, has described the affected region as The Bermuda Triangle—an area bounded by Miami, Bermuda, and Puerto Rico. The accompanying map shows the areas of the southwest North Atlantic Ocean where, in the last century, ships, planes, and boats have unexplainably disappeared. The majority of these disappearances were probably caused by vent tunnel emissions; however, over the centuries, the unexplainable disappearance of some vessels may be attributed to storms, collisions at sea, poor seamanship, defective hulls, defective machinery, faulty construction, faulty designs, accidents aboard such as explosions and fires, hijackers, acts of barratry, assailing thieves, mutinous crews, navigational errors, incompetent crews, and attacks by drug runners, among other reasons.

You will see on the map that a preponderance of losses occurred off Florida's northeast coast, and was not spread uniformly between Miami, Bermuda, and Puerto Rico. Few disappearances occur near Bermuda and Puerto Rico. The vast majority occurred off Florida's east coast, and this continues to be true to this day.

In his book, "Into the Bermuda Triangle" (International Marine / McGraw Hill, 2005), Gian J. Quasar states that from 1980 to 2005, over 1,000 boats and planes disappeared. And he lists them at the rear of his book.

The Florida/Bahamas area is in line with the area where the asteroid buried itself, but the waters here are relatively shallow, thus shortening the distance the erupting material has to climb in order to affect boats and planes in the area. Elsewhere in the so-called "Bermuda Triangle," the sea is much deeper because it is east of the continental shelf. Therefore, a more appropriate moniker might be "The Florida Shelf."

In December 1945, five TBM Avenger torpedo bomber planes, comprising flight 19, were sent on a short military training mission over the Bahamas and back. They left from Fort Lauderdale, Florida and were never seen again. A big PBM Mariner plane with a crew of thirteen was sent out to try and find the planes of flight 19, or wreckage, or bodies, or survivors. This plane also disappeared and never returned. Later an underwater search was made, and it located several TBM Avenger aircraft on the seabed off Fort Lauderdale. It appeared that flight 19 was found. But the serial numbers found on these planes did not match those of flight 19. And the mystery continues.

Certainly not all ships that have disappeared in the southwest North Atlantic met their fate in a manner similar to the loss of the "Atlantic Bulk Princess," described later in this book. Still many marine and aircraft disappearances have not been satisfactorily explained, and have turned into the stuff of legends.

Chapter Fifteen

Bubbles

I remember when I was about two or three years old, dreading a certain time of day: bath time. The idea of getting water all over my body was repugnant to me. But Mom always won the battle, and my screaming and hollering did me no good. She eased the trauma by adding plenty of bubbles, which made the water fun to splash around in. I enjoyed my bubble bath once I got wet, but the apprehension just before was terrible.

Then one day Dad came home with a present for me: a small wooden tug boat, painted red with a black smoke stack on top. The boat was made from a flat piece of pine with a smaller piece glued on top to represent the superstructure. It was about eight inches long and had shoe eyelets for portholes. The first time I played with it in the tub, I placed it on top of the soapsuds and to my horror it disappeared. My hand shot down after it, and I found it floating on the water below the bubbles. After that, Mom stopped putting suds in the tub anymore, and I began looking forward to bath time so I could play with my boat.

The point of all this is that boats (of any size) don't float on soapsuds or bubbles. Only a boat that is lighter than air—like a blimp—can float on suds. But no one has ever made such a boat, obviously. A boat, or a large ship, is designed to float in a medium of water, which has the proper density needed to support a hull.

On the other hand, a submarine can flood its ballast tanks with sea water to increase the vessel's weight, enabling it to sink in the water. It can then partially blow out the tanks when necessary, while moderating its hull speed and dive planes to level off under water. To surface, it simply increases the hull's buoyancy by blowing more water out of the ballast tanks, reducing the hull weight and enabling the submarine to rise to the surface. Even a submarine would sink in a medium of bubbles. The bubbles started out as trapped sea water, sitting in the Earth's magma along with remains of the asteroid. As explained earlier, the bubbles broke through the seabed and escaped as steam, rising to the surface of the sea.

These crustal vent tunnels can be of any size, from a few inches to hundreds of feet in diameter. One surviving example that comes to mind is on land on the island of Bimini in the Bahama Islands. It is now a small, deep, freshwater pond. Its width narrows as it winds its way down until it disappears far below the surface. Because this tunnel broke through on dry land, it never closed up again because there was no great depth of water over it to collapse and close the tunnel.

Another that may or may not have been a series of tunnels is Lake Okeechobee in South Florida. This is now a nearly circular, shallow, freshwater lake, and is the second largest lake in the U.S. (The largest is Lake Michigan. The other Great Lakes don't count—they are half in Canada). Okeechobee may be a very early vent tunnel, and is now a shallow lake filled with silt and sediment from fresh water rivers feeding it over the centuries.

Revisiting Armageddon: Asteroids in the Gulf of Mexico

The fact that it is nearly circular leads me to suspect it was formed as a result of an early crustal vent tunnel—or a concentration of many crustal vent tunnels—or possibly an ancient asteroid impact crater. The fact that it is round lends plausibility to this theory.

Other examples include Deep Lake, a freshwater hole on the north side of US Route 29, north of the Tamiami Trail in Collier County, Florida. Silver Springs in central Florida could be of similar origin.

It is possible that the salt water that started out in these holes eventually drained off, leached down through the limestone formations, and was replaced by fresh water draining down from the north as time went by. These concepts require further research, but maybe someday people much smarter than this writer will prove or disprove what I say.

Chapter Sixteen

Rolling Limestone

The Florida peninsula, especially the southern portion, is covered by a layer of limestone. It would appear that this limestone was built up long before the Gulf Event when the area was inundated to a shallow depth by the Atlantic Ocean. Because in south Florida this limestone lies on the surface of the land in many places, especially in the Everglades, it is readily accessible, and has thus become a source of fill, rock and sand to use for commercial purposes.

Surface excavators of various types dig out huge chunks of limestone to put in hoppers which grind the rock into smaller uniform sizes, or gravel and sand.

In other places around the U.S., limestone and other types of rock are found in solid layers that require breaking up in order to remove it.

South Florida limestone is different. Much of it is already severely cracked and subsequently the edges have become cracked and broken. The Gulf Asteroid is the obvious reason for this. When it landed in the Gulf, it buried itself in Earth's mantle as it crossed under Florida and then under the seabed off Florida's east shore, where it finally became lodged partially in the magma and partially

Revisiting Armageddon: Asteroids in the Gulf of Mexico

in the crust.

As it passed under Florida, it created a rolling wave as it first raised the land surface and then immediately dropped it back in place, as it passed by.

The solid layer of limestone was cracked in the process. Some pieces can be held in the hand; others would fill a wheelbarrow, and others are as big as a small car.

It reminds me of a similar ripple effect when the fans do "the wave" at Fenway Park in Boston. People in center field suddenly stand up, raise their arms, yell, and sit down. Then people to their left do likewise, followed by fans in right field, along the first base line, behind home plate, along the third base line, left field, and out to the foul pole next to the Green Monster. Perhaps you have observed or participated in this procedure in a sports stadium somewhere.

Sometimes things in nature don't work to perfection; for example, in Florida, when the land was raised, it didn't fall back into its original form. The resulting landscape features small areas of land which are a few inches higher than the adjacent areas. On these "islands" known as "hammocks," which appear haphazardly all over the Everglades, plants and grasses grow and eventually trees which needed dryer surfaces for their roots grow there as well.

Conversely, this wave effect left some places in lower positions than they had been before. These areas were flushed out with rainwater after the Gulf Event and became shallow ponds. In periods of heavy rain the ponds would overflow and follow the lowest path south to the Gulf of Florida. These paths became permanent ditches, rivulets, streams, and "traces." None contained any fast moving water because the land was too flat to allow it. These watery paths are still seen today all over the Everglades, where developers with bulldozers are not yet allowed to venture.

MYSTERIES OFF FLORIDA'S SHORES

For hundreds of years, ships, boats, and more recently, airplanes have encountered strange anomalies and unexplained forces in the southwestern part of the North Atlantic Ocean region. Many have survived by overcoming or skirting the problem or by reversing course. Others, in their opinions, were just plain lucky. Many other ships, boats, and planes have disappeared entirely, crews included. A few large ships have been located and explored by divers and submersibles.

No plausible explanations for these sinkings have ever resulted, based on the limited examinations of the sunken wrecks. Furthermore, we should not fault the divers and crews of the submersibles; they just didn't know what to look for, as you will learn later.

When I refer to "hundreds of years" above, we must remember that no ships ventured into open areas of the southwestern North Atlantic prior to about 1000 AD, when Viking explorers from Greenland explored our North American shores from Baffin Island to Cape Cod, the Florida peninsula, the Gulf of Mexico, up the Mississippi River, up its tributaries, and southerly along the Gulf coast in Mexico. Other than the open seas between Greenland and Baffin Island, the Vikings stuck close to the North American eastern shoreline. They had no reason to venture easterly or southeasterly into the unknown.

After the Gulf Event occurred, the first ship to sail into this offshore area was Columbus in 1492 AD followed by many other

Revisiting Armageddon: Asteroids in the Gulf of Mexico

European explorers in ensuing years. But the sea conditions capable of creating the mysteries mentioned above trace their beginnings to the Gulf Event about 7,500 years ago. In the beginning, immediately after the Gulf Event, sea conditions were ripe to cause problems for just about every ship in the area, but of course, there were no vessels out there for almost 7,000 years.

As centuries rolled by, these mysterious sea conditions recurred less and less frequently until now they are quite sporadic and localized.

It is not every month that we hear of ships and planes encountering problems or disappearing in those areas mentioned. Certainly, boat and plane manufacturers in the southeast U.S. do not want problems and disappearances getting media attention—it's bad for business. But the U.S. Coast Guard has records that show alarmingly high numbers of sea vessels and aircrafts that go missing every year. We cannot assume that these losses won't happen again, every year, for the foreseeable future. They will continue into the future, but at a very slowly decreasing frequency as time goes by. Certainly these problems at and above sea level will continue for many centuries to come, until sometime in the distant future the sea bed will settle down and go into a permanent sleep to some unknown degree.

The next four chapters will give you an idea of what happens when a ship is caught up in a fatal sea disruption. The names of people, ships, and most of the places are fictional and have no bearing on anything past or present, living or dead. But the story is my best shot at telling you as best I can what probably occurred aboard a hypothetical ship.

Ray Covill

CHAPTER SEVENTEEN

A Great Ship Goes Missing

Rio Grande Energy, LLC was owned by Japanese interests. They were deeply involved in energy development, both oil and gas, as well as extracting and moving ores around the East Coast of the United States. The company had been headquartered in Houston, Texas, since the mid-sixties. One of their efforts was running a shipping line to haul bauxite and other ores from Texas through the Chesapeake to the docks in Baltimore.

RGE chairman, Mitsui Bishimoto, decided in the mid-1970s that there was more money waiting to be made by acquiring another coastal ore carrier. When he wanted something, his staff jumped and rarely questioned his judgment. His managers in Houston started the search for a yard that could deliver a new ship by Bishimoto's deadline: July 1, 1977, as it was known around RGE, "L Day," for launch date.

Over the years, Bishimoto's marine development manager had dealt mostly with yards in Japan, Holland, and Germany because of sound business relationships and favorable exchange rates for the US dollar. Unfortunately, at this time the yards in Yokohama and other Japanese cities were booked for the next three years; the Dutch were also jammed up, and the yards in Germany were interested but their schedule was too tight. The boss distrusted the yards in Hong Kong, and ultimately the contract was awarded to the Maine Metalcrafters Corp., an American company. Most of their work was for the US Navy, building new ships, and renovating, refitting, and retro-fitting ships.

Maine Metalcrafters had a ship for the Navy already under construction, with a stretch job on a Naval Supply Ship to follow on railway #2. But Navy contracts, for the usual budget and political reasons, were becoming increasingly difficult to land, and there wasn't much on the horizon, and so the deal was made between Maine Metalcrafters and RGE.

The day eventually came when the new Navy ship under construction slid down the ways into the river. Within the hour, Maine Metalcrafters's giant crane was lowering heavy steel beams onto the now empty railway for the hull framing crew to weld into a keel for Bishimoto's new pride and joy.

He had already picked out a name for her: the "Atlantic Bulk Princess." Other crews in the yard started building various pre-assembled sections for the ship which would eventually be lowered into place, then welded and bolted together. When all sections were in place, the time consuming work of assembling and installing the HVAC, wiring and electronics, plumbing, and painting followed.

Almost everyone in Portland, where Maine Metalcrafters was located, was dependent, either directly or indirectly, on the shipyard. As its fortunes went, so did the city's. The ironworkers union and other unions got along like brothers with Maine Metalcrafters' managers, knowing their mutual successes were dependent on each other. Maine Metalcrafters had never failed to launch a new ship on schedule, even if this meant putting on a second or even a third shift to meet deadlines. Maine Metalcrafters, the unions, and the workers all knew if a ship was launched past the scheduled date, financial penalties would arise, eat into their profit-sharing, and shrink the new money going into their 401K's and bonuses.

On the other hand, everyone also knew that if the ship was launched ahead of schedule, large credits for each day would not only benefit the Maine Metalcrafters, but would also filter down to benefit every worker at the yard. When Maine Metalcrafters' engineers and marine architects worked up a bid, they knew roughly how long it would take them, and how long it would take the other yards that were bidding the job, even after allowing the usual "fudge factors" for weather, strikes, change orders, late material deliveries, etc. The cooperation and harmony between Maine Metalcrafters and its unions was unique in the industry, as was the yard's reputation for excellent workmanship, and especially Maine Metalcrafters uncanny ability to deliver ahead of schedule. Those Yankees at Maine Metalcrafters knew how to make money. The yard had an enviable track record for successfully constructing many ships for the Navy over the years.

As the months rolled by, construction progressed normally. By February 1977, the engineers calculated that the job was going so well that they could beat the L-Day schedule of July 1 by forty-one days. Bishimoto's engineers from Houston had been paying regular visits to Maine Metalcrafters to check on progress, work out minor problems that arose, approve substitutions and change orders.

By April 15, Bishimoto's top two engineers were living

Revisiting Armageddon: Asteroids in the Gulf of Mexico

near the shipyard and working with the Maine Metalcrafters team on a daily basis. Bishimoto was thrilled with the yard's progress and lack of problems. L-Day was moved up to the third Friday of May.

When launch day arrived, all of the shipyard's workers, their families, and all the city's school children were invited, since the schools always closed if a launch day fell on a school day. Most of the businesses shut down for a half day, too, to attend the launch.

The actual launch had to occur within fifteen minutes of high water, when the tides on the river would be slack. The extra depth was also needed to prevent the ship from grounding in the mud after her slide into the water. The tide would be high at 7:03 a.m., so the launch was set for 6:48 a.m.

The speeches started at 6:15 with the CEO of Maine Metalcrafters, followed by four union heads, the State Representative, and the Senator for the region, the Mayor and then Bishimoto himself. The speeches were running behind schedule, so the Maine Metalcrafters boss told Bishimoto to keep it down to 8 ¼ minutes, which he did.

Then his wife broke the 1.5 liter bottle of Japan's best sake over the ship's bow, thereby christening the ship, and down the ways she slid. None of that champagne nonsense for him! Sake brought better luck.

The waiting tugs nosed up to the ship, stopped her northern momentum, and then pushed her gently against the dock where deck hands threw the lines to waiting hands on the dock, who placed the hawser loops around the huge bollards. Another successful launch had gone off like clockwork.

Cars on the nearby bridge had all stopped and blew their horns. Pleasure boats nearby in the river blew theirs as well, and throughout the city, church bells rang, old people clapped, others yelled their approval, and the kids screamed with delight. The band music was drowned out completely. All the VIPs retreated to the tent erected in the east parking lot for an elegant breakfast hosted by Bishimoto. It was a joyous day.

The final fitting out, as well as testing of electronics and all the assemblies were done dockside over the next month. Bishimoto's wife did not like the white paint color in the ship's mess hall and wheelhouse, and so they were repainted baby blue. She too had her input.

Bishimoto was anxious to get the ship into operation and thus insisted that sea trials be conducted while en route to Texas instead of returning to the shipyard. If any problems occurred they would solve them while underway. The workers and technicians aboard, if all went well, would be put ashore in Boston, and if more time was needed, put ashore in Brooklyn and flown back home, courtesy of Bishimoto. Any major problems could be resolved in the Brooklyn yards, where he had made arrangements for such an eventuality.

But all went well, and the sea trial people were disembarked in Boston's outer harbor, one mile east of Boston Light. The Atlantic Bulk Princess was built to the latest Lloyd's standards, based on her class, type, and tonnage. She was registered out of Monrovia, Liberia. The ship had an LOA of 586 feet, beam of 78 feet, and was designed to hold a cruising speed while fully loaded of 16 ½ knots. Her maximum speed was 17 ½ knots.

Revisiting Armageddon: Asteroids in the Gulf of Mexico

Bishimoto's people in Houston had selected a crew of experienced Japanese men who all spoke English and had served on other company vessels. But the job of captain created a new problem. There were no available Japanese candidates within the company because the fleet had increased over the last five years and highly reliable, experienced top officers were scarce.

They decided to give the captain's job to a fifty-eight-year-old Norwegian who had served the company for seventeen years, had twenty-six years of experience as a master, and who was just returning from a leave of absence. The captain would be Anders Eieslund. Bishimoto knew him well and had complete confidence in him. Eieslund insisted that his crew call him "Cap'n Hank." He was stern, but fair, placed the safety of his crew first, was an expert mariner, and his crew felt it was an honor to be aboard his ship. They loved and respected him, and some of them even looked up to him as their father; all of them trusted him enough to follow him to the ends of the earth.

Eieslund had spent his early years near Oslo and served on his father's fishing boat from the time he was fourteen. His family came to the United States in 1955 and settled in New Bedford, Massachusetts, where they had relatives active in the scallop fishery. Eieslund persuaded his parents to let him enter the Massachusetts Maritime Academy in Buzzards Bay. He had his sights set on something beyond a crew member on a scalloper or dragger. His parents agreed and he graduated in 1960 with honors, and then spent two years with other shipping lines on coastal freighters. He then found a job with Rio Grande Energy, learned to speak Japanese, and the rest is history.

By late July the Atlantic Bulk Princess was in Houston, and began to take on a full load of bauxite bound for Baltimore.

Cap'n Hank watched the loading operation from his place in the wheelhouse. The stevedore gang worked quickly and efficiently as the conveyors swung to fill every nook of the ship's twelve huge cargo holds.

The ship was designed with six holds running fore and aft down the starboard half of the deck, and six more on the port side. Between each group of four was a ship's crane, used to remove and replace the hatches, swing the conveyor belts, and move the front-end loaders in and out. The cranes were a new swivel-type design that could rotate 360°. A display panel in the wheelhouse showed twelve green lights to indicate which hatches were secure, and twelve red lights to indicate if any were ajar or removed, or otherwise not battened down securely. It was helpful for the captain to know if a hatch cover had loosened during a period of heavy weather, when the seas were crashing over the deck. He could change course and speed and take evasive action to minimize the stress on a hatch until the crew was able to get out there and re-secure the hatch after the weather improved. This could prevent a hold from flooding and causing more problems.

When the last hold was loaded, the ship's three cranes picked up the hatch covers and lowered them into place for the stevedores to secure. Captain Eieslund decided to get underway early in the morning, not only to make the harbor pilot's job easier, getting the ship from the docks in Houston down the channel and into the Gulf, but to make it safer to get through the usual congestion offshore near the entrance channel and to skirt around the assortment of drilling platforms scattered around the captain's intended course in the northwestern Gulf.

All went well, and when the last drill rig was safely astern or well abeam to the north, the captain increased the hull speed to the designed 16 ½ knots.

At 6:00 p.m. he placed his scheduled call to Houston to report on the ship's performance, location, exchange any messages, and get an up to date weather report from Houston.

He was very pleased with the way the ship was handling. All was well as they steered a course to take them south of the shoals between the Dry Tortugas and Key West. They intended to more or less follow the 40 fathom line south of Florida Bay and east of the Florida Keys. As they did so, the course slowly turned northward into the Florida Straits toward the Gulf Stream between Miami and the Bahamas.

At about 10:30 a.m. on the third day out, the ship was about forty miles ESE of Marathon Key and slowly turning northward, making her usual 16 ½ knots. The weather was perfect—81°, no clouds, visibility clear to the horizon, and an easterly wind of ten-to-twelve knots, barely enough to create occasional whitecaps.

The captain was in his compartment directly aft of the wheelhouse, nursing a cup of coffee, and looking over a chart of the Straits. First Mate Yoshiro Sakaturi, 28, was at the wheel. He was held in very high regard by Eieslund.

At 10:34 Sakaturi noticed a low fog bank dead ahead about three or four miles away. It was very difficult to estimate the proximity of a fog bank because there is nothing solid to look at and nothing adjacent to compare it to. He signaled the engine room to reduce speed to seven knots, turn on the ship's running lights, and decided to call the captain, which turned out to be unnecessary because as soon as Eieslund heard the engine slow down he was out of his room and on his way to the wheel with Sakaturi.

"Why did you cut engine speed? Is anything wrong?"

The first mate showed him the fog bank ahead. A deck officer added, "Radar shows nothing near us—a full clear horizon."

Eieslund took out his binoculars and could see nothing but fog ahead. "What's NOAA giving for weather?" The first mate checked and said to the captain, "Weather for the straits calls for S/E, ten to fifteen, seas under three feet, visibility unlimited. No mention of fog."

The captain then took the radio mike, went to channel thirteen (the ship to ship channel), and tried to contact any other vessels transiting the Straits. No reply. He then went to channel 21A (a United States Coast Guard channel), and called USCG Station Key West. They came right back and said they had no other reports of fog in the Straits area, and suggested it may be just an isolated patch.

The captain asked if there were any reports of a line squall moving through the area. Again, a negative reply from Key West. Cap'n Hank had been through more than a few line squalls before—the last one aboard a New Bedford scalloper on Vineyard Sound near Gay Head—which looks like a low, black fog bank and moves rapidly. Line squall winds can elevate to hurricane force in two minutes, blow for ten minutes, and then die down, only to be followed by similar winds from the opposite direction. Then the squall disappears as quickly as it came. The greatest danger is to small boats, but not the Atlantic Bulk Princess. She could handle such a thing in stride.

"This doesn't look like a line squall," Cap'n Hank said to Yogi, his nickname for Yoshiro. "And besides, they are very rare in this area. It's not moving fast, if at all, and the color is wrong. This looks more like a putrid greenish yellow than black."

The captain reduced speed to four knots and set the fog horn to automatic. "If this damn fog bank is bigger than I think it is, it could delay our scheduled arrival in Baltimore, and that's not good. Stay alert Yogi and let me know if you see anything. Keep an eye on that radar too. We don't need a collision at sea; that could ruin our whole day," he said laughingly. There was no need for undue concern.

By 10:45 a.m., the ship was soon entering the fog, visibility dropped to a quarter mile, and then quickly to one hundred yards. The captain noted that this was very unusual for the Florida Straits. He could barely make out the silhouette of the high bow superstructure. He looked at the compass to check on his heading. It showed southwest. On second glance, the compass began to move counterclockwise to southeast and then back again before going into erratic short swings in all directions. Eieslund had never seen a compass act this way before. *Maybe it'll settle down*, he thought to himself. *I can still see the sun off the starboard bow, nearly abeam, and high in the sky.* He knew his course was still northward.

10:46—At this point the Captain was concerned but not duly alarmed. The hull seemed to vibrate more than what he expected at four knots. He had the third mate call down to the engine room to talk to the chief engineer.

10:48—The mate looked worried. "Cap'n Hank, Chief says the shaft is turning faster than it should for the throttle setting he's using. It's getting hot in the engine room. Instead of the usual 83°, it's now 102° and rising, and everything is soaking wet from the humidity. What's going on Cap'n?"

"Well I'd like to know too. Something's screwy." The captain checked the ship's speed log. It read only one knot through the water.

"Yogi, what's GPS read for speed over the bottom?"

"It says zero, Cap'n Hank."

"How the hell can that be? Check the backup, Yogi, and while you're at it, check the LORAN too."

"Both say we're dead in the water Cap'n."

Cap'n Hank checked the compass again and found it still spinning. Then he opened the cover on his other magnetic compass, ahead of the main one on the dash. It too was spinning erratically. Puzzled, he next took a reading from the gyro compass and this one indicated North by East, the course he thought he was on, which relaxed him a bit.

10:48—Eieslund was once again becoming very concerned. The crew members in the wheelhouse could detect it in his voice, and they were more than concerned—more like frightened. They looked to their captain for guidance and leadership. If anyone could get them out of this mess it would be him; there was no one else to turn to anyhow. There was only God, and more than one of the crew was sending out silent prayers in His direction.

10:49—The captain opened the starboard door leading out to the wing and was hit by an intense heat. He choked on the hot, foggy air which was acrid, pungent, and stunk like burning sulfur. He held his handkerchief up to his mouth, and looked aft at the ship's wake. The screw was churning up froth under the fantail. But something was wrong; the wake did not trail away aft of the ship. In fact, there was no wake at all. He then looked forward to check his bow wake. There was no "bone in her teeth"—no wake here either.

10:50—Eieslund went back inside the wheelhouse to discuss the situation with his first mate. "Yogi, I don't know what the hell's going on, but we're going to get out of this." He called down to the engine room. "Stop engine. Give me full astern." The engine slowed, and then started up again. The hull shook like never before.

Yogi watched both GPS instruments and the LORAN. "Cap'n we're still dead in the water. I think I know what the problem could be. We can't make any headway 'cause we're hard aground on some uncharted reef."

"No way, Yogi. The Straits are deep and free of any shallow water. Look at the Fathometer. How much water we got?"

"Must be a mud bottom, Cap'n. The echo is coming back weak and waivers between forty-five and fifty fathom."

"See what I mean, Yogi? Think of something better."

10:52—"Put her in forward gear and give me max power."

"Aye, Cap'n," came the reply.

Eieslund swung the wheel to full right rudder, and verified it with the rudder position indicator. His hands were locked like vices on the wheel. The rudder was full right, hard up against the rudder chocks. The ship shook under the strain from the engine.

"Yogi, what are we making over the ground?"

"Still zero knots, Cap'n Hank."

"What's wrong with this damn thing?" Eieslund replied. He looked out at the sea. It was flat. The gentle easterly swells they saw all morning had disappeared and so had the wind. He looked again at the sea and saw it covered with bubbles on both sides of the ship. Eieslund suddenly realized that the sea was boiling—literally boiling.

"Yogi, ever heard of any volcanoes in these waters?"

"No, Cap'n, the nearest one is two-thousand miles away in Southern Mexico, and that one is pretty dormant, I think."

"Look at the sea, Yogi. Am I nuts or is the damn sea boiling?"

"Cap'n, you're right. What's going on?"

"Wait a minute, Yogi; look at the starboard side of the deck. I see green water slopping onto the deck." Eieslund ran out onto the starboard wing and looked at the side of the hull up forward of amidships. The sea was lapping over the starboard rail. Her normal freeboard when loaded was about twenty-one feet, and it had disappeared. He ran back inside and said "Yogi, we're sinking—get out a May Day so they'll get our position. Make sure you have fed all our data into the radios. At least they'll know our name, P.O.B.'s (number of persons aboard), and Lat/Long. They'll know where to look for us, and put that May Day on channel 13, channel 16 (calling and emergencies), channel 21A and 22 (Coast Guard frequencies), and channels 68 through 72 (pleasure boat frequencies). And while you're at it, put it out on single sideband. That carries all over the place. Hop to it."

The captain grabbed the phone and called down to the

engine room, as well as crew's quarters and mess. Before he could speak, the chief engineer started yelling to the captain, "We can't stand it any longer. Air temp of 121° and rising, the outer hull plates are too hot to touch, the engine is working, but acting up, and everything is soaking wet and hazy here."

The captain interrupted him and said, "Give me max rpm's."

"But sir, that'll put it over the redline."

"Screw the redline; do it. Lock the throttle wide open, then grab your boys and get up here on the double!"

The chief didn't even think about what he had been told; he just started moving. The men started the scramble up to the bridge deck.

10:55—When they arrived in the wheelhouse, the men looked half dead: no shirts on, soaking wet, reddened skin, half delirious.

"Yogi, are you getting any responses on the radio?"

"No, Captain, nothing on any frequency. All I'm getting is steady static."

Eieslund took the ship's mike in hand, "Attention all hands, attention all hands, this is your captain speaking. All hands to the wheelhouse on the double. We're going to abandon ship." He then checked the hatch indicator panel again. Twelve green lights, meaning not a drop of water was getting through the hatches. Then he looked at the twelve automatic bilge pump lights which were also all green. The bilges were dry. He wondered, *How the hell can we be sinking if we maintained the hull's integrity and we're not taking on any water? It makes no sense.*

10:59—Frightened crewmen wearing their life vests were now in the wheelhouse and soon all souls were accounted for.

"Boys, we're going down, and I can't explain why. We've got May Days going out on all three radios and the *Emergency Positioning System* is already overboard. When you hit the water your EPIRB will pop off and start signaling immediately. When your boat has settled down, crank the motor and steer west by north. We're about fifteen miles off the central Florida Keys. Pick out a sandy beach and I'll see you all at station Miami. Good Luck. Now get the hell outta here. You too, Yogi, get moving."

"No, sir, I'm staying here with you. No matter what. You'll need me to help with the GPS, LORAN, radios, and whatever else. I'm staying here."

"You've got a wife and three little kids back home. Get the hell out of here."

"No way, Cap'n, I'm here to stay with you and that's final."

It was useless to argue so Eieslund let him stay.

11:02—"Yogi, take the wheel. I'm going aft to make sure the boys get in the water okay."

Eieslund followed the men to the top of the slide where they were climbing into the aft hold of the lifeboat. The ship's lifeboat was a twenty-four foot hard plastic V-bottom with hard chines to minimize rolling, and a ten-foot beam. It was perched on a pair of rails at the same level as the bridge, and slanting down at a forty-five degree angle to a point above the railing at the stern. It could either be lowered manually over the stern by the stern crane, or in an emergency the brake could be released so that it would slide down the rails, over the stern, and plunge bow first into the sea. It was extremely buoyant and would pop up to the surface after its plunge and float like a cork.

The lifeboat contained twenty-four extra life vests strapped to the ceiling above the bench seats and a small electric motor near the rudder. In the tiny superstructure at the stern which rose above deck level, a man operated a folding tiller. Sitting on a shelf a foot above the keel, ten huge, deep cycle batteries, which also served as ballast, were strung from bow to stern. They could power the boat, depending on sea conditions, at four knots for up to a week, or two knots for up to a month. There was a small wired generator atop a short mast above the superstructure which helped to recharge the boat's batteries, as long as the wind was blowing.

The lifeboat had running lights, stern and bow hatches, a signaling light, a mirror, a whistle, a rope ladder and a boathook. On board, there was enough food and water for several weeks—or longer if the rations were smaller, two knives, toilet paper, a first aid kit, fishing equipment, flashlights, interior lighting, a manual horn, two air-powered horns, two decks of playing cards, reading materials and a VHF radio. Two collapsible buckets served distinct purposes: one to collect rainwater and one to be used as a toilet. For navigation purposes, there were coastal charts, a compass, and an EPIRB in case of emergency. Additionally, a four-foot sea anchor on 150 feet of poly line would keep the bow into the wind in rough weather to ease the ride for the crew inside. The boat was also equipped with a lifeline all around the waterline, a telescoping mast and boom, an orange plastic sail that could be placed through the hatch in a bracket in order to sail the boat, a hinged dagger board that could be lowered next to the keel, a 6-foot-square piece of plastic, an electric bilge pump with a manual backup, a bailing scoop, and hammocks with blankets tied against each side of the boat.

The outer hull was painted a bright orange. It also had reflective paint in certain areas in addition to the reflective tape, flares, radar reflector, orange smoke sticks, and two flare guns. It had six portholes on each side that could be opened in good weather and an electric vent system with a manual backup. The lifeboat was simply the best money could buy. It had bow and stern cleats and a two-hundred-foot towline.

After the last crewman had climbed in, he secured the hatch, belted himself in, unlocked the release handle, pulled down the brake, and the boat shot down the rails into the sea.

As the captain watched, the boat popped up after the impact, and then slowly started to settle, bow first into the sea where it spiraled down. Within ten seconds, only the top of the stern hatch was visible, and then that too disappeared. The captain could not believe what he was seeing. It was impossible. That boat had to float.

Aboard the lifeboat, the men, all wearing their life jackets and belted into their seats, felt the splash and waited for the boat to right itself and assume a level keel position. Once this happened, they would start the motor and head for the Keys. But the boat didn't right itself; instead, it held its twenty degree down angle. Daylight coming through the hatches slowly grew dimmer. Men began to panic, unbelted themselves, and started to try and get out through the hatches while other men on board tried to restrain them. Complete chaos ensued as the frantic men screamed in panic. They were sinking; they could not believe what was happening.

Within a minute, they were sixty feet down. Suddenly the bow hatch imploded with a bang and the sea roared in. The aft hatch blew out and water was everywhere. No air spaces were left. Most of the men drowned where they sat. One got loose and tried to climb out, but he was dragged back into the cabin by the suction force of the water. Within minutes all the men aboard the lifeboat were dead. Their plastic coffin continued its slow, bow down spiral until it hit bottom at 400 feet, and came to its place of eternal rest.

The EPIRB on the lifeboat could not float either. It shot to the sea bottom long before the lifeboat did.

11:06—The captain ran back into the wheelhouse. He decided not to tell Yogi what he had just seen. It could serve no constructive purpose. There was no point in alarming him any further.

"Cap'n Hank, I'm glad you're back. We got green water half way up the cranes. I can't see the bow anymore. What's happening, Captain? We're sinking. Are we gonna die? I don't want to die." He kept repeating it over and over.

Panic set in for both of them. Yogi was crying and sobbing hysterically.

The captain stood at the wheel, his hands white at the knuckles from his grip on the wheel. With his eyes transfixed dead ahead, he waited for his ship to respond to the helm. Then he heard an explosion far below the wheelhouse and the ship started to shake. The tachometer needle fell off to zero. He knew what had happened: the boiler had exploded and the engine had seized and shut down.

Yogi continued his hysterical sobbing. He stood behind the captain, bracing his back against the rear bulkhead, slowly filling his pants.

Eieslund stared straight ahead. Water was now at window level in front of him. He knew they were going down, and there was absolutely nothing he could do about it. A surge of helplessness overcame him as he thought of his wife and family. Would they ever know what had happened to him? He began to pray out loud. It was getting dim inside the wheelhouse. He instinctively turned on the emergency lights and turned the horn switch to continuous blow.

11:08—Suddenly the starboard door leading to the bridge wing extension imploded. The seas blasted in. Eieslund's ears popped and then the portside door exploded out. Yogi was swept up by the surge of water and thrown head first toward the port doorway. On the way through, his head slammed against the top of the door frame. He was blown out the doorway feet first, his neck broken. Yogi was dead. His limp body shot out into the hot, stinking murk, and disappeared forever. He was the only crewman aboard the ship who would never have to endure the horrible agony of death by drowning.

The force of the water could not rip Eieslund's hands free from the wheel and he drowned there, eyes staring straight ahead, hands clenched on the wheel in an eternal death grip. It would take weeks for the normal rotting process to allow his fingers to release their grip on the ship's wheel. Only then would his soul be able to rest. He had given the effort his best shot, but failed.

The ship slid gently beneath the waves. Her horn gurgled

to a halt. The entire ordeal starting from the time the bow poked into the fog bank had taken about twenty minutes. She settled, still upright, with a slight bow down angle. When she reached the bottom, the bow hit first, breaking the keel amidships, followed a few seconds later by the stern crashing into hardpan. Clouds of mud and silt encased the ship, but soon cleared as the gentle bottom currents took over. The ship had found her final resting place, where she would sit for centuries until rust and deterioration had eaten into her enough to break what was left of her into pieces, and litter the ocean bottom with chunks of her steel frames.

Davey Jones had claimed another prize. The Atlantic Bulk Princess, her cargo, and all of her crew… were gone.

Port View of a Lifeboat Freighter

Chapter Eighteen

Transit Airlines Flight 981

An assortment of people were crowded into the gate area still waiting to hear boarding instructions for Transit Air Flight 981, from Port of Spain, Trinidad, non-stop to Miami International Airport. Its scheduled departure was to be at 8:56 a.m., but it was running a little late.

When the plane lifted off the runway at 9:21, all the deadpan and grouchy faces were now smiling, and strangers were talking to each other. The captain came on the intercom and apologized, "Problems in landing on schedule in Miami which were beyond our control..." he explained. He went on to say they expected to arrive in Miami almost on time—at about 12:05 p.m. rather than 11:56 a.m. That was not enough of a delay to foul up the connecting flights for people.

At 11:25 a.m., the aircraft reached a position north of Cuba and about sixty miles east of Marathon Key. Miami Traffic Control called 981 and had the flight captain descend to 5,000 feet, vector right, west of Abaco, into a counter-clockwise pattern to land on runway number one. The pilot repeated the instructions back to traffic control before making the necessary changes. This new flight path took them on a northerly course up the middle of the

Revisiting Armageddon: Asteroids in the Gulf of Mexico

Florida Straits at 5,000 feet. They would soon get further instructions to drop lower, and line up on runway number one.

The pilot noticed that some of his instruments were erratic so he and the co-pilot began to run a test program schedule. The compass was slowly turning first in one direction and then another. Fortunately, the skies were clear and the pilot could see the upper Keys and knew for sure he was on course in spite of what his compass indicated. Suddenly, his altimeter reading dropped to show 1,000 feet. Again, the pilot did not panic since he knew the plane was still at about 5,000 feet just by looking at the sea surface and islands.

He decided to report his instrument malfunctions to ATC Miami just in case, but now his radio wasn't working. He could only get static on all frequencies. A bit of consternation in the cockpit resulted, but Transit Air's pilots were too well trained to panic.

At 11:40 a.m., ATC Miami called 981 to give the flight captains final landing instructions but got no answer. Worse yet, 981 suddenly disappeared from Miami radar.

The air traffic controller working 981 called the supervisor over, and standard emergency procedures went into effect. All departing flights were held back, and all incoming traffic was stacked above 15,000 feet in a holding pattern, or else diverted to other airports. All planes in a 250 mile radius were requested to report any sightings that appeared to be a four engine passenger jet. Crash crews were alerted to standby status in case they were needed. At 11:54 a.m., ATC was becoming concerned that they had lost 981 and that she had gone down.

The United States Coast Guard at Station Miami was put

on alert and planned to begin a search by 12:30 if communications with the plane were not re-established.

At 11:42 a.m., 981 tried to call Miami but the radio was still out. The pilot decided to ball park the last instructions and got ready to make a visual approach to land whether or not his onboard capabilities were working. He knew ATC would have him on radar and would give him a clear shot to land on runway number one.

At 12:20, 981 suddenly showed up on ATC's radar and radio communications were established once again. A sigh of relief overcame the ATC people in the Miami tower. 981's instruments were back to normal also. The landing went routinely after that. The plane was brought into a gate in a relatively remote area used on occasion by a few international cargo flights.

Miami authorities questioned everyone aboard 981. The pilot reported the cockpit problems with systems acting irregularly until they finally heard from ATC with landing instructions. The cockpit dash clock and watches worn by nearly every person aboard read 12:08, the pilot's approximated time for gate arrival.

Eastern Standard Time in Miami was, in fact, 12:37. Where was 981 for the missing twenty-nine minutes? Why was there this discrepancy? Why did they disappear from radar? Why did communications disappear and then reappear?

The passengers were interviewed and then sent on their way. The plane was towed to an empty hangar and quarantined. Transit Airlines experts started their investigation and the airplane manufacturer sent a team to help out. Then the National Transportation Safety Board (NTSB) headquarters in Washington sent a delegation to investigate. The NTSB decided to let the investigation proceed and not notify the media until they could

Revisiting Armageddon: Asteroids in the Gulf of Mexico

come up with an explanation that the public might accept so everyone could put the entire matter to rest.

That effort failed before it even began when a number of passengers called newspapers and TV stations to report that all of their watches were twenty-nine minutes off. A media feeding frenzy ensued when reporters, especially TV reporters, jumped all over the story like a pack of lions after a crippled antelope.

NTSB forbade any of the investigators from talking to the press. At 8:00 that evening an NTSB spokesperson in Washington held a press conference where he read a prepared statement saying in summary that the incident was under investigation, that their staff was working with the USCG Station Miami, and a further report would be made public when the investigation was completed. He would not take any questions.

The media was furious, as expected. It could take years to conclude such an investigation and by then the story would no longer be hot or even relevant. The TV and radio stations interviewed the public, held polls, and acted in their normal fashion.

The poor, gullible public ate up the incident programming which suggested that the Atlantic Bulk Princes and Flight 981 had both gotten caught up in the Bermuda Triangle. Other reports implied a UFO had been the cause of the missing ship and flight interruption. TV polls showed that viewership had skyrocketed, as expected. The media loved it. They were going to reap rich rewards.

One little old lady named Hortense Bottomsteiner, from Lumbago, Idaho, made national TV with the statement that two weeks ago her left knee started to ache and that was a signal that she was having a vision. This vision told her that Flight 981 was going to disappear and that the people behind it were "little green men from Mars."

The NTSB and all others who investigated the incident couldn't come up with any sort of explanation. They disassembled parts of the plane and went through it, testing and closely examining everything. The plane was eventually put back into service and functioned normally.

The government officially closed the case, unbeknownst to the public. As months and years rolled by, people eventually lost interest and the entire matter was successfully buried in the government's files someplace, never again to be resurrected.

Mayday

At 11:15 a.m. on the day the ship disappeared, J.L. Bennett was about to shut down his ham radio station so he could help his wife with some chores in the kitchen. He was one of the most prominent ham operators on the East Coast. He terminated his last conversation with a fellow operator in Phoenix, Arizona, but then decided to see what was on CW. He scanned the frequencies and picked up a faint, feeble SOS. His mind instinctively copied any Morse code coming through. He heard the dit, dit, dit, dah, dah, dah, dit, dit, dit only once. He tried to fine-tune the receiver squelch by tweaking it a bit off the static frequencies with a little success. He then copied "Princess" followed shortly after by "Strait," and that was the last he heard.

Bennett tried to put two and two together. This signal was probably coming over the CW sky wave and not by ground wave. If it was by ground wave the source would be within fifty-to one-hundred miles maximum, and "5 by 5" (meaning loud and clear).

Revisiting Armageddon: Asteroids in the Gulf of Mexico

Sky wave signals could be coming from anywhere on the globe and didn't always come through as clearly.

He contacted a few other operators from Massachusetts to Florida but no one else had picked up on the S.O.S. Bennett's buddy in Miami suggested that the call may have come from a ship in distress in the Florida Straits because there were no other prominent "Straits" anywhere in the region.

At 12:25 p.m. or so he asked his Miami friend to notify USCG Station Miami on the long shot that it could be something named "Princess" calling from the Florida Straits. Miami thanked him for the report, made an official record of it, and tried to call "Princess" to no avail.

With no location to go by, and no other information to help them, the Coast Guard sent out a PAN alert to vessels east of Miami and south as far as Key West, to keep a lookout for any vessel in distress, any vessel with the name "Princess," and to report any sightings or contacts to Station Miami, or any other Coast Guard Station. Soon the media learned of the search for a vessel in distress and began their own investigation.

STRANGE HORN SIGNAL

Earlier that day, at around 11:00 a.m., Lee R. Austin and his wife were sailing their chartered Tartan 38 from the Dry Tortugas back to Marathon Key via the Florida Straits when they head a deep horn signal. They were both excellent, experienced blue water sailors, having sailed together for thirty-nine years in the Florida waters, the Bahamas, and four times in the British Virgin Islands. The signal they'd heard had to be coming from a large ship. Lee looked around to make sure he was well clear of any commercial traffic. He was sailing northerly, about two miles east of the Keys.

At 11:20, Lee heard the ship's horn go on continuous blow, and thought it very unusual. He glanced at his watch and made a mental note of the time. Why would a horn signal go to full blow? He had no answer, but did not give it too much thought. He was almost up to Marathon Key and ready to start his downwind approach around the outer entrance buoy into the channel leading to the marina.

A Search Begins

At 10:00 a.m. EST that day, A.J. Dupont, Houston's radio officer, was awaiting the scheduled daily call from the Atlantic Bulk Princess. No call came. At 10:05 he called the ship. No reply. Dupont continued to call every five minutes until 11:30 but still he got no response. Finally he picked up the phone and called his boss, Tajimoro Seki, Vice President of Coastal Operations, who was surprised at the lack of contact. Eieslund was as reliable as a clock. Seki asked Dupont to continue calling the ship until 12:00 p.m. He then checked his weather reports and learned that a high pressure system extended from the Texas shore to the outer Bahamas. So he at least knew that the failure to establish radio communications was not due to poor weather conditions.

Based on the ship's location at 11:00 a.m. the previous day, Seki calculated that the ship should be in the Florida Straits somewhere east of the central Keys. He then called USCG Station Miami on the land line to report the problem and ask them to try and contact the ship.

The Coast Guard updated Seki on the Transit Air Flight #981 incident as well as the ham operator report, and told Seki that a search was to begin shortly. Seki then alerted the rest of the

Revisiting Armageddon: Asteroids in the Gulf of Mexico

management team in Houston, and called Bishimoto in Tokyo.

Bishimoto was at home when Seki reached him and he was shocked by what Seki told him. He couldn't believe a word he was hearing and he had a hundred questions, none of which Seki could answer.

Stations Miami and Key West each dispatched three forty-foot patrol boats. Miami then alerted all ships at sea in the area to report any unusual sightings, including any debris in the water.

The officer in charge at Station Miami called another hurried meeting of his staff, including search and rescue people. They now decided to order their two large, long-range search planes into action as soon as the crews were aboard and the engines warmed up. Air Traffic Control at Miami stopped all traffic from landing or taking off until the CG planes were up and clear of the immediate area.

Secondly, the Coast Guard decided, reluctantly, to notify the media. A few pieces of the puzzle began to fall into place. But there were still questions that could not be answered with certainty: Did the disappearance of the ship have anything to do with the airplane incident? And what about the May Day report from an operator in New Jersey?

Within an hour after the story hit the local TV channel, Lee Austin decided to call Station Miami to report his hearing an unusual ship horn signal. The Coast Guard felt that was a key bit of information, because it was the first clue as to an approximate location of the ship.

Coast Guard Stations Miami, Key Largo and Key West sent three additional boats to conduct the search effort, plus two

CG Cutters to search around the clock.

Four days later they stopped sending out the forty footers and one week later they recalled the big ships. The planes searched in daylight for four days and none of them found a thing, not even debris. The PAN alert bulletin that had been sent out to ships at sea also came up with nothing.

The Atlantic Bulk Princess had in fact disappeared. The NTSB had some of their people at Station Miami. The station commander shut and locked the eight-foot chain link fence gates to the complex to keep the media hordes at bay so the station could maintain some semblance of sanity as they continued their business and try to find the missing ship.

The entrance area in front of the gates was clogged with six TV trucks, their squiggly antennas poking high above the commotion. Reporters, TV crews with cameras on shoulders, and curious onlookers were crowded against the gates and adjacent fences, staring at the CG building and wondering what they were looking for. Miami police had a difficult time keeping the entrance drive clear of trucks and people to allow official CG vehicles to enter and leave the station. The station's officer on duty saw the circus outside and asked the station commander if it would be a good idea to send someone out with a statement for the media, in an attempt to placate them, at least temporarily, and get rid of the crowds. The OIC wrote out a statement and a midshipman took it out to deliver.

The midshipman got within ten feet of the gate and decided he would go no closer. He read the statement which said, in summary, that the ship Atlantic Bulk Princess was not at the location they expected her to be, and that a search has commenced,

Revisiting Armageddon: Asteroids in the Gulf of Mexico

and when further information became available, the press would be notified in due course.

Media reporters had their cameramen focused on the midshipman as he read from the report and once he'd finished, they immediately began yelling questions at him. No one could be clearly heard as they all called out simultaneously.

Finally a well-dressed woman in her forties came up to the gate with a microphone and a cord leading to a TV truck. Her name was Margaret Ambrosia, but she like to be addressed as "Ms. A." She was 41, 5 feet 4 inches tall, divorced with no children, and she was the Chief Operations Officer (COO) of a major national news bureau's Miami office. Her staff, behind her, referred to her as "Coo Coo." Very few of them held her with any degree of esteem, mostly because she had a terrible hang-up about her height. She refused to hire anyone, male or female, over 5 feet 6 inches which prompted competing news media people referred to her news team as the "runt squad."

Ms. A. was convinced the top jobs in the American industry were all held by men ranging from six feet to seven feet tall and she was determined to make an inroad at any cost. This is why she dressed like a man—in pant suits. She also favored the pants to hide her funny looking, skinny legs. All of her pant suits—five in all, one for each day of the week—were tailored in various shades of black. She had tiny feet, always perched precariously upon six-inch heels which her staff referred to as "railroad spikes." Her ankles, weak at best, were frequently sprained from falls she incurred while wobbling around, until she finally hit on the idea of buying wide-width shoes to give her more stability. This is how she acquired her other nickname, "fat feet."

Still feeling inferior to men, she always had her hair done up in a tightly pointed bun so that her head looked like an upside down ice cream cone. From the tips of her six-inch heels to the top of her pointed pile of hair, she gained several inches and measured in at around six feet tall.

The loudspeaker system on Ms. A.'s truck drowned out everything else. She demanded entrance to the station and demanded a full disclosure from the "admiral in charge," from whom she wanted answers to all her questions—immediately.

She further informed the poor midshipman who was standing, spellbound and frozen, that this Coast Guard building was public property, that she and her TV station paid a lot of taxes and that they, therefore, had a right to get into the station any time they wanted. Furthermore, if he didn't unlock the gate, she "would report him to the Pentagon."

When the station commander heard this through his open, second floor window, he sent two more midshipmen down to grab the first one and drag or carry him if necessary back inside the station, which they did.

The police, with reinforcements, dispersed the crowd and sent the TV trucks and staff on their way. The people screamed and hollered at the police about their rights as citizens to enter public places, and that they have a duty to inform the public what is going on, blah, blah, blah.

The press had to make do with whatever information they had, and then interview all kinds of prominent people: political personalities, anchor persons, and instant "experts" to glean opinions from them about their interpretations of the recent events, and what they thought was the probable cause.

Revisiting Armageddon: Asteroids in the Gulf of Mexico

Hortense Bottomsteiner, the little old lady from Lumbago, Idaho, made another national TV appearance, and said again the cause of the recent events was "little green men from Mars." Her eighty-seven-year-old husband, Egbert, stood behind her giving the cameras a show of his big, toothless grin. Today was their second claim to fame.

While all this was going on, Bishimoto and his two top people were boarding the company's private jet in Tokyo en route to Washington, D.C. Once at cruising altitude he phoned the NTSB to alert them about his expected arrival time at Dulles Airport. He wanted an immediate meeting with USCG and NTSB officials.

A limousine picked up the Bishimoto party at Dulles and took them directly to the meeting in Washington. They had brought with them a suitcase full of naval architect's blueprints, designs, work drawings, photographs, and schematics for the U.S. experts to study and see if they could find any design flaws that may have led to the ship's demise. They found nothing; there was absolutely nothing suspicious about any of the architectural plans.

Bishimoto next asked the U.S. Navy to intercede and search the bottom of the suspected area with side scanning sonar in hopes of finding the ship on the assumption it may have sunk. By now most everyone was resigned to the belief the ship was down somewhere in the area east and north of Florida.

With the help of the Japanese Ambassador, some strings were pulled in the right places and one navy ship from Miami and one from Guantanamo got underway to begin the search of the Straits of Florida.

On the second day, the U.S. Navy located an uncharted wreck on the sea bottom, with an approximate length that matched the size of the Atlantic Bulk Princess. The navy felt it was still premature to commence a deep water examination, but with some inducement payments to the right people, plans moved forward.

The navy had no deep water submersibles available on the East Coast at that time. Bishimoto contacted the Woods Hole Oceanographic Institution in Woods Hole, Massachusetts, and after some negotiations, hired WHOI to send one of their oceanographic research vessels with the deep sea submersible "Alvin" aboard, to the Florida Straits. It took five days for the WHOI people to prepare Alvin and its mother ship and get to USCG Station Miami.

The weather forecast for the Straits was favorable, so the ship proceeded to the site early the next morning and Alvin went to work. Within hours, the two men aboard found a ship in their flood lights right where the navy sonar reports had placed it.

They slowly circled the ship until they passed by the stern. There, clear as a bell, they read "Atlantic Bulk Princess-Monrovia." Moving slowly around and over the ship they could see that her hull plates amidships on both sides were broken open which they correctly assumed was caused by her collision with the bottom at the end of her bow first plunge. There was no debris field scattered around on the seabed. Her screw appeared intact and her rudder was hard over on full right position. There was no evidence whatsoever to indicate a collision at sea had occurred. A close look at the aft superstructure showed that many porthole windows were broken and they could see what looked like a body inside the wheelhouse at the helm. They also noticed that the ship's lifeboat was missing.

What they could not see was that the main boiler in the

Revisiting Armageddon: Asteroids in the Gulf of Mexico

engine room was blown apart, the throttle locked on full ahead position, and the fact that the engine had overheated and seized tight. Nor did they see that the cutlass bearings around the shaft had melted or come apart from the high RPM and hot sea temperature, and that the hatch covers were tightly secured.

Several weeks later the final reports on the sinking attributed the loss to "undetermined causes." The media even hinted at a cover-up, and that the truth was being hidden. Just what kind of cover-up, by whom, and for what reasons, they did not say, but it made good fodder for the gullible public to ponder.

The expected public polls on "what caused the sinking" ensued, with no conclusive results, and all the television news programs and newspaper editorial pages allowed all the political and media people to become instant "experts" and voice their opinions and conclusions. The public gobbled it up.

As weeks and months rolled by, the sinking slowly took a back seat to countless items of more immediate interest to the public, until eventually the matter became just another footnote in history.

Ray Covill

CHAPTER NINETEEN

Why the Ship Sank

The Atlantic Bulk Princess sank because it entered a large area of live steam bubbles pouring out of many crustal vent tunnels spread over a concentrated area. Her propeller could not push or pull her out of the area because the propeller simply spun around among the bubbles, and thus could not generate any lateral force to move the hull. Because there was no movement over the bottom, the rudder was useless.

The medium of foam and bubbles was not dense enough to permit the propeller to take a "bite" in it. It simply sloshed around in this atmosphere of bubbles. The longer the hull sat in the bubbles, the more it developed negative buoyancy, and slowly settled until she finally slipped beneath the waves. This result would be the same for a boat of any size, from a small pleasure boat to a large freighter or tanker. They are all designed and built to effectively operate in a medium of water. Anything less viscous won't work.

NAVIGATION SYSTEM GONE AWRY

To understand the failure of the compass, LORAN, radio equipment, EPIRB, and other systems aboard the hypothetical ship and any crafts above it, such as Transit Air Flight #981, we must first look at some basics.

I believe that sometime in the first billion years of the earth's formation, a massive asteroid, made mostly of iron, landed in the area we now know as the Arctic Ocean, north of Canada. It is not located at the true north geographic pole. That would have been too much of a coincidence.

It was large enough, shallow enough, and concentrated enough to disperse magnetic radiation around the globe. This allowed the later development of the compass, which indicates the direction of the magnetic north pole, thus permitting safe navigation over Earth's surface.

On a much smaller scale, the Gulf Event produced something similar, although vastly inferior.

Going back to the story of the ship gone missing, the burst of live steam that shot out of crustal vent tunnels from Earth's interior were made up not only of sea water but also various gases such as sulfur dioxide, and more importantly, the molecules of steam were impregnated with particles of electro-magnetic radiation from the Earth's hot magma.

This was caused by several unique features that were present at the birth of the conversion of the sea water into steam. First, the water was surrounded by the molten iron from the asteroid as well as magma from Earth's interior. Secondly, it was placed under incredible pressure from the weight of Earth's mantle and sea water above it. These factors combined to burn electro-magnetic radiation into every molecule of water. The heat and pressure were so great that the water could not expand into live steam. It stayed in its original form as a liquid until it broke free of the seabed.

When the water escaped via the crustal vent tunnels it temporarily forced open, it turned into steam as it shot out, creating streams of bubbles, and then formed gaseous "fog banks" above the surface of the sea in that immediate area until the pressure under the sea floor diminished enough to allow the sea weight to close up the tunnel. These irregular and concentrated bursts of steam then "confused" the ship's magnetic compass needles causing them to change direction for very short instances. The compass aboard the ship, in this manner, rotated in both directions and spun wildly in other instances. In addition to these rapidly changing signals, the fathometer and other electronic-based navigational devices had to contend with the normal magnetic signals from Magnetic North, which came into play when the signals from the vent tunnels weakened between bursts at intermittent short intervals.

Magnetic forces cannot be detected by humans. But birds, for example, do tap into Earth's magnetic force to help them find, and stay on their migration routes over Earth's surface. The same holds true for turtles, whales, some fish, butterflies, and other living creatures.

The radiation from the vent tunnels can extend well above the "fog bank" up into the stratosphere, especially if the steam escaped from a great number of tunnels over a large area of the sea, thus affecting compasses, radio signals, altimeters, and other electronic devices in airplanes that may be located in the immediate area. It is very doubtful that a space vehicle in orbit above the Earth would feel any of these effects because instead of measuring elevation in feet or meters, they may be in orbit many miles above the surface—far too distant to be affected by signals from the seabed.

Revisiting Armageddon: Asteroids in the Gulf of Mexico

Once a plane had passed through the affected area, the compass and other electronic instrumentation would suddenly return to normal, and in many instances the pilot would have maintained his speed, elevation, and direction if instrument confusion did not panic the pilot.

Conversely, if a pilot panicked and began to believe some of the readings on his panel, it would be very easy to quickly lose control and consequently, the plane would easily go down at sea. For instance, if a pilot had been flying at 2,000 feet and his altimeter suddenly showed 11,000 feet, he may believe the instrument and drop down 9,000 feet to his planned cruising altitude, causing him to crash into the sea, especially if he was offshore, at night, in poor visibility, in cloud cover, or the sea surface was flat and difficult to see. Another example would be if the pilot could not see any sun or stars to refer to for navigation, and instead of heading for land, he were to fly in circles or trust his compass and go northeast further out to sea until he ran out of fuel, perhaps hundreds of miles from shore. One can see how a tragedy could easily occur.

Could something like the above examples have been the explanation as to what happened to Flight 19 when five planes and one search and rescue aircraft went missing off Florida's north coast in December 1945? No one knows for certain unless a successful search is made well northeast of Florida, into the deep water off the edge of the continental shelf. But nothing is guaranteed.

Strong emissions of electro-magnetic radiation, in large quantities, can not only make a compass useless as indicated above, but the field of scatter emanating from the live steam can completely blanket and neutralize radio wave signals moving through the affected area, such as AM, FM, VHF, UHF, CW, microwave, radar, sonar, fathometers, LORAN, EPIRBS, GPS, etc. The only exception I can think of would be for a boater to check water depth via a hand-held weighted line. But this is useless in heavy seas and deep water. For all intents and purposes, there are no alternatives to the functions if a pilot or sailor is well offshore and cannot see any sun or stars.

To illustrate this phenomenon, tune an AM portable radio to a weak frequency between signals so it is picking up mostly background static. Turn up the volume. Then take it near an operating microwave oven or better yet, a hair dryer. The closer you get, the louder the static noise becomes until the radio signal is completely drowned out when the radio is held close to the appliance.

CHAPTER TWENTY

Time Shunts

First let me say it is a known fact, (rarely discussed or researched), that intense fields of electro-magnetic energy can alter the normal passage of time as we know it. Back in the mid-twentieth century, The U.S. Army got wind of an experiment the Nazis worked with in World War II in Germany. Not much is known about the results of their efforts.

The U.S. Government allegedly tried an experiment to observe how intense fields of electro-magnetic energy can affect the passage of time for animals and objects in the grid work of the magnetic force created by the energy. If it is true, in October 1943 the U.S. Naval ship USS Eldridge was docked at the Philadelphia Naval Shipyard. The government laid many lengths of cable around the perimeter of the main deck. With crew aboard, huge amounts of electric current were run through the cable. The crew suffered horrible deaths and disappeared, along with the ship itself. The disaster was well covered up by the government, and not much information has been pieced together since the disaster occurred. Numerous sources have questioned the truth of this experiment.

In the case of hypothetical Transit Air Flight #981, referred to earlier, the people aboard the plane had no knowledge about the distortion of time that occurred when they were in the vicinity of the Atlantic Bulk Princess disaster. Their apparent time loss did not become known to them until they landed at Miami International Airport. Then the media circus took over.

I cannot adequately explain the details of the matter. But time losses, and in some cases time gains, have been known to occur sporadically for over the last fifty years, at least. They undoubtedly occurred long before the middle of the twentieth century, as far back as Biblical times.

Could Methuselah have really lived for 900 years? In terms of apparent Earth time, I think that is quite impossible, unless a "year" to people of that era was a month as we know it. Or did some controlled force, electromagnetic or otherwise, allow him to live for short intervals on Earth with his peers, and disappear for longer intervals in a ratio of about twenty to one, meaning one short span of time on Earth and twenty years somewhere else, in which apparent time to him may have seemed like just one year. If he could have worn a watch during his entire lifetime, one that measured not only minutes and hours, but years and centuries, his apparent breathable years may have totaled only forty or fifty, but to people on Earth, his elapsed time here, including the time when he was AWOL, may well have been 900 years.

Just how he could have engineered this loss, by finding a place of controllable electromagnetic energy (which is highly unlikely), or due to the interaction of some other force, I do not know, and I hate to speculate on it. It would be interesting to try and conjure up plausible scenarios that could lead to fascinating assumptions and conclusions, even to the point of trying to put myself in his shoes and assume he had no knowledge or control of the time changes he endured.

The more I tend to allow my mind to focus on any given theoretical subject matter, and then simply let it run riot, the more plausible scenarios start to fit together, all into place, and then lead to other related situations and all the possible solutions that ensue.

In my case, it takes concentration over long periods of time with no interruptions. Pure day dreaming is usually simply the re-hashing of past experiences into current scenarios, but what I describe above is something quite different because it leads to virgin areas and new ideas.

If Methuselah's life was stretched to 900 Earth years, but only forty to fifty "breathable years" this constitutes a phenomenon I call an "accelerating time shunt." Time sped up for him. Exactly how and why, I do not know.

If this should some day in the future be explained by scientists, then there is probably also a reciprocal phenomenon I call a "decelerating time shunt," where someone's body ages more rapidly than what would have been normally experienced in terms of Earth years. It could be only a few weeks or months, or many years.

Before this new century has run its course, I think we'll know a lot more about these matters, and why our government kept us in the dark and failed to take any action. I'm sure it will be shocking and angering. Our government, for at least 100 years, in many areas, tells us only what they want us to know, and they simply disavow all else and answer no questions. Eventually, time buries everything. I think there is a good chance we will learn more about these matters before the 21st century ends. We'll just have to wait and see.

But getting back to Transit Air Flight #981, the intense electromagnetic energy that broke free through the earth's crust off the Florida coast permeated not only the sea, but the atmosphere above, and was the cause of the twenty-nine minute time loss. The amount of energy, meaning the force of energy, and its sustained levels (be they steady or rapidly peaking and subsiding), resulted in a decelerating time shunt of a small magnitude—in this case twenty-nine minutes.

It is highly unlikely that information about matters such as this, for example, will ever be retrievable through the Government Free Public Information Act. It will be categorized as just another matter that is not recoverable by us since it is deemed to be in the "best national interest" to do so. And there it will end.

But then again, maybe not. Once our government has figured out how to create and control this time/energy continuum, probably late in this century, then it will be only a short matter of time before foreign governments in Europe and Asia will use their spy network to steal our secrets. I base this statement on the track record our government has created in the last sixty-five years or so. Consider how short a period of time we were able to keep secret our information about the atom bomb, rocket technology, intelligence networks, the recent "Wiki-leaks" disaster etc., and you will know what I mean. It's all driven by the allure of quick and easy money and international science competition between nations. Nothing is going to change that, I'm sorry to say.

Similarly, newly developed sciences regarding: 1. The temporary reduction of the mass of a tangible object, (including humans), to zero; and 2. The then accompanying science relating to crossing of Earth's force lines to create controllable, ultra high speed motion, will also suffer the same fate as the time/energy matter referred to above. Again, only time will tell.

CHAPTER TWENTY-ONE

The Formation of Florida

Once the asteroid struck in the Gulf of Mexico and caused a piece of the Atlantic Ocean to move rapidly ENE, evidence does exist, believe it or not, to show the exact direction and also pinpoint where the central area of the flood struck in western Europe and on the northwest shoulder of North Africa.

The layout of the Florida peninsula runs roughly from northwest to southeast. It was formed by the fallout in the region immediately in front of the impact area. It is safe to assume that most of the debris landed directly in front of the blast center and lesser amounts landed off to the sides. The central part of the peninsula has a bulge on the western side about half way down the peninsula which would indicate that this was the central landing area, having received more debris than other parts of the peninsula. This area is now called Tampa—St. Petersburg.

Opposite this, more or less, on Florida's east coast, is another bulge on the shoreline, which we call Cape Canaveral. The Cape, however, is not connected to the mainland; it is a wide barrier island, and the only one of its kind on Florida's east coast. The rest of the islands are long narrow barrier islands formed by sand being pulled from deep under the length of the islands and built up in storms. The sand built on the islands until they assumed their current shapes.

Storms coming up the coastline continually move beach sand further up to the north along the beaches, and at the same time continue barrier building onto and over the beach lines.

Cape Canaveral could have started out as a debris pile from the central area of the Gulf Event fallout centered a few miles further off shore than the rest of the east coast. Then, as storms from the east pounded the shore, they would have struck this Canaveral area head on and dumped sand over and beyond it to the north, leaving a large shallow area of sand behind.

As the speed of the storm waves traveled over the Cape, they would have slowed down with no solid wall of water behind them to keep them moving rapidly. This "fallout" and deposit area of sea-borne sand and gravel thus left behind a large shallow area to the north of Canaveral.

If you travel from Titusville, Florida across the inland sound and over to the north end of the Canaveral area, you will see an area of many square miles of ponds, mud flats, and salt water grasses, and some dry sand extending well north of Canaveral. This is now called the Merritt Island National Wildlife Refuge, and coincidentally is one of the best birding spots in Florida.

Revisiting Armageddon: Asteroids in the Gulf of Mexico

Deep Lake - Route 29, Jerome, Florida

    The edge of the continental shelf from the Florida Keys to the Cape Canaveral area comes closer to the mainland than anywhere else on the United States East Coast. The Gulf Stream comes through this area and is close to the mainland because it has to squeeze between the Keys and the westernmost Bahamas.

    As the shelf continues north it veers further out to sea, so that from Cape Canaveral to Virginia it is many miles further offshore. From Virginia to Cape Cod it continues even further offshore. The area off Florida's east coast and out to the edge of the continental shelf is called the Blake Plateau, a fairly flat area with depths ranging up to 3,900 feet. Beyond this point, the waters of the Hatteras Abyssal Plain range out toward Bermuda with depths reaching up to 17,000 feet in places.

    The bulge on Florida's west coast falls in the latitude of St. Petersburg. Moving eastward from here to find a place that approximates the center of Florida in its longitudinal axis shows that the town of Gibsonia, Florida (east of Tampa), is the general area of the center of the fallout area in front of the asteroid impact zone.

The distance from east of Marathon Key to Gibsonia is around 200 miles. This would define the approximate southern half of the immediate fallout area of the asteroid. Now if we look 200 miles or so north of this center, we find that the northern terminus lies in Georgia, near the Okeefenokee Swamp half way from Fargo, Georgia to Homerville, Georgia and east of Route 441.

The ocean waters rushed in to fill the void caused by the asteroid and would have sloshed around in many directions until settling down and assuming its current level and new shorelines. This "sloshing around" of the water polished the land and accounts for the reason Florida is so flat today, especially the southern half.

Chapter Twenty-Two

Across the Pond

The next thing that comes to mind is just where this slice of the Atlantic Ocean struck land in the European/North African region after it was set in motion by the asteroid. There is an easy way to solve this riddle and the solution starts back in Florida. To understand the trajectory of the debris post-impact, on a globe draw a light pencil line down the Florida peninsula from the Fargo, Georgia terminus through the Gibsonia area, and down to central Marathon Key, more or less. Now place the short edge of a piece of paper, 8" x 10", or larger, along this line. Slide it along the line until one corner is on Gibsonia. Now draw a line along the long edge from Gibsonia out into the Atlantic Ocean. Extend it with other sheets if necessary along the same path and it should find land around Morocco on the northern shoulder of Africa.

If you do this on a map, it will come out in Europe some place because the Great Circle Effect will be missing, and the higher the latitude, the more it will exaggerate the true path of the slice of ocean as it sped to the northeast. For this reason, you will obtain a more accurate path if you follow the above instructions on a world globe instead of a map.

Since water is fluid, even though it is much denser than air, it can easily move when pushed. Each molecule presses against the next one on the lee side until the pressure ceases and then friction and gravity bring the motion to a halt.

The Gulf Asteroid push sent the sea charging forward for almost a thousand miles while the asteroid tore into the earth's crust where the Florida peninsula now sits. By the time the asteroid had buried itself deep in the earth, water and earth debris were still on their skyward leg, and then came down to cover the peninsula as described earlier.

The push of the asteroid obviously blew a lot of sea water up into the stratosphere ahead of it, but it also pushed water off to the sides. This explosion took place in the Gulf, but its impact traveled all the way to North Africa by way of the sea. The sea at the explosion site was sent traveling as a 3,000 mile-long slice of the sea all the way into North Africa and Southern Europe.

Chapter Twenty-Three

Northern Flank of the Flood

As the great wave approached land in the North African and Mediterranean areas, it would be more appropriately referred to as the "Great Flood." So, from now on, I will refer to it as the Great Flood, or just the Flood. Floods do not occur at sea, of course. The term "flood" refers to water inundating land somewhere.

First, let's take a look at the Mediterranean Basin. It is, of course, a wide depression in the earth formed millions—maybe billions—of years ago in the formative years of the planet, when the then massive singular land mass was created. It appears that several monstrous asteroids struck in the Mediterranean Basin area between Gibraltar and Turkey. I say more than one piece—maybe many—because of the elliptical shape of the basin. One big piece would have created a more circular pattern.

It is entirely probable that the Black Sea and Caspian Sea were also created at the same time. They would have been the result of smaller asteroid impacts than the one described above.

The torrent of asteroid impacts created depressions in the earth's crust from Gibraltar to The Caspian Sea. The general direction of rock masses that struck were headed in a northerly direction and at an angle of up to 60° which is fairly steep. Such an angle is evident to me because there are no mountain ranges in the southerly part of the basin, meaning North Africa, from Tunisia east to the mountains of Turkey.

Going clockwise from Algeria, there are other mountains on the westerly part of the basin. They are the Pyrenees Mountains in northern Spain, the Alps in northern Italy and beyond, the mountains in the Balkans on the northeast side and the mountains in Turkey to the east.

The string of mountains we call the Apennine mountain range in the Italian peninsula, plus Sicily and Malta, were created very shortly after the other ranges mentioned above, and it was all due to a different phenomenon than what is described above.

When an asteroid strikes Earth at a fairly steep angle and on mostly dry land, here is what occurs, in the simplest of terms: the initial contact results in an explosion that generates great heat and power. The materials comprising Earth's cover at that point are immediately compacted and super-heated, resulting in a large dust cloud that spreads to the immediate distribution area, with larger pieces of earth and rock blown the furthest outward.

The dust is then carried by wind patterns all over the hemisphere—or in some cases, all over the planet.

The diameter of the impact area's hole is in direct ratio to the asteroid's size, composition, and speed upon impact. The floor of the impact crater is the point where no further material in the hole is blown out. The floor is compacted downward (in the case of a smaller impact), or it is depressed and followed by the asteroid plunging into Earth's mantle, in some cases all the way into the interior molten magma.

Revisiting Armageddon: Asteroids in the Gulf of Mexico

In cases where the floor survives and is not created because the asteroid goes into or through the mantle, a type of knee-jerk reaction occurs. The severe pressure of compaction which ceases almost right away, allows the jammed-down Earth to sort of "bounce back" upwardly, somewhere in its central location.

This can be seen in the Yucatan, Mexico asteroid impact area which occurred about 75 million years ago, the dust from which caused the dinosaurs on Earth to starve and become exterminated. A rise of land in the central area of this crater has survived.

To give you an idea of how this works, drop a pebble into a dish of water with a flat calm surface. First, the pebble causes a splash on all sides as it goes into the water. The water level under the pebble is pushed downward for an instant. Then, when the pebble is underwater the surface of the direct area of compact rises up into the air, and is then pulled down by gravity, and the water surface returns to its flat "normal" configuration. The concentric waves sent out in a circle around the impact point would, of course, take a bit longer to subside.

I don't think this process can be demonstrated on dry land. It needs an asteroid to create the effect because dry land is not fluid.

If the effect is extremely severe, as in the formation of the Mediterranean depression, the "bounce-back" reaction described above can display itself in a larger, more severe result.

The compression in these cases could open up weaknesses in the crust or even leave passageways open over a large area. This then allows the interior magma to rise and pour out onto Earth's surface, creating enormous volcanic action in the form of flowing magma and explosions. The result is large hills or mountains. This is how the Apennine Mountain range, running down the Italian peninsula to Sicily and Malta, was formed. You will note that this north to south split is about half way between Gibraltar and Turkey which is where we would expect it to occur.

The time it took for the creation of all the above effects, meaning depressions and mountain ranges, must have taken thousands of years—maybe millions of years. Indeed, it is still going on at a snail's pace within Mt. Vesuvius and Mt. Aetna in Italy, as well as through earthquake activity that still continues to present times in Turkey, northern Iran, and that general area.

If all the above current activity is due to an asteroid impact that happened millions of years ago, or maybe even longer, then it will probably continue for thousands of years into the future. Someday it will eventually subside.

The human race may have been around for many thousands of years, but even if we talk in terms of several million years, it is still the blink of an eye on the asteroidal time continuum, and there is thus no reason at all to assume that asteroid impacts, tsunamis, and volcanic activity are anywhere near the end of their life cycle.

CHAPTER TWENTY-FOUR

The Drowning of a Basin

Before the water reached the North African area, it first encountered some groups of islands west of Portugal and Morocco in the eastern Atlantic Ocean, including the Cape Verde, Azores, the Canary Islands and Atlantis.

The wave, while west of these islands, slowed a bit due to the friction of shallower waters. It rose in height and smothered the islands leaving deep water on the west sides and shallower water on the east sides, as a result of dropping suspended silt and sand on the lee side of the islands.

It is entirely possible that these islands were connected by land to the African mainland, but there is no way to know for sure, except for the "near" connection that is discussed later on in this story when I discuss the matter of Atlantis.

As the water neared the coast line of the African and European continents, the water again grew shallower and therefore slowed a bit, causing the front of the wave to simply rise upwards from the seabed and crash over the land. Its height by that time could have been miles high, dwarfing anything similar in recorded history.

The flood raced into the Mediterranean plains, an area we

now call the Mediterranean Sea. But prior to this happening, the basin was a landlocked, huge, fertile valley with rivers, lakes, and undoubtedly inhabited by a large human population.

The Straits of Gibraltar did not exist prior to the Gulf Event. They should more properly be referred to as the Isthmus of Gibraltar, a low land running several hundred miles from west to east, and connecting Spain to Morocco.

Since gravity allows water to move more rapidly through areas lower than adjacent higher areas, the speed of the water over this low isthmus scoured it out into a strait fairly quickly. Due to the fact that the water was moving rapidly, it tended to hold the sand, mud, silt and gravel it dug out and kept it in suspension until the water found some higher land in its path. As it passed over higher land, such as the mountains and hills along the Italian peninsula, the lower sections of water then slowed and dropped out much of the material they were carrying.

The bedrock formations protruding out of the Straits of Gibraltar withstood the onslaught, but lost their mantle of earth, gravel, sand, and stones because all this material was instantly washed away eastward, denuding the rock formations into what we see there today.

The slice of Atlantic Ocean that did all this did not keep its width (north to south) intact like a laser beam does. It mushroomed out as it moved easterly and slowed, growing most likely into the shape of an exponential horn. The majority of this water wall moved into the Mediterranean plains and North Africa, but very significant amounts were peeled away to the north and less to the south of the center.

Since the wave struck North Africa at a shallow angle, the

southern flank rolled into and over the more southerly side of the North African shoulder with less notable effects than the northern flank did, because its northern angle was closer to the general direction of the main stream.

It is quite conceivable to me that the northern "overflow" could easily have cleaned off the lower Iberian Peninsula, the Cherbourg Peninsula, and the southern coast of England, leaving these west facing shores with deep water close to land and rocky shorelines

The sea went over the low land bridge area between England and France, then picked up speed as it went through and scoured out the Dover Straits. This could explain how the rolling English hills at Dover were chewed away and became the White Cliffs of Dover. The south shores of Ireland may also have been eroded but probably much less than England and Cherbourg, because Ireland lies more to the west and is off the beaten track a bit, so to speak.

This northerly "overflow" branch subsided as it entered the North Sea, flooding the shallow area of Europe opposite England, thus dropping plenty of material which had been suspended in the fast moving water, into this region. This low, or "nether" land area, we now call Holland.

It has long been believed by scholars that a land bridge between England and France once existed, but just how far back in time has been anyone's guess. At least up until now that is.

Chapter Twenty-Five

The Mediterranean Basin

The sea we now call the Mediterranean Sea occupies a depression created in Earth's formative years— four to five billion years ago. The skies back then were black, ultra hot, covered with volcanic soot, ash, and dust with enormous amounts of solid materials in orbit.

Gravitational forces slowly strengthened as more and more solid matter was pulled in toward the center of the congealing, rapidly growing mass, soon to be a planet.

When a surface began to form around the planet, Earth's interior materials kept the central core in a semi-liquid state which is still the case today. But the thin, outer layer cooled enough to form solid land. The seas formed later on when the surface of the planet had substantially cooled.

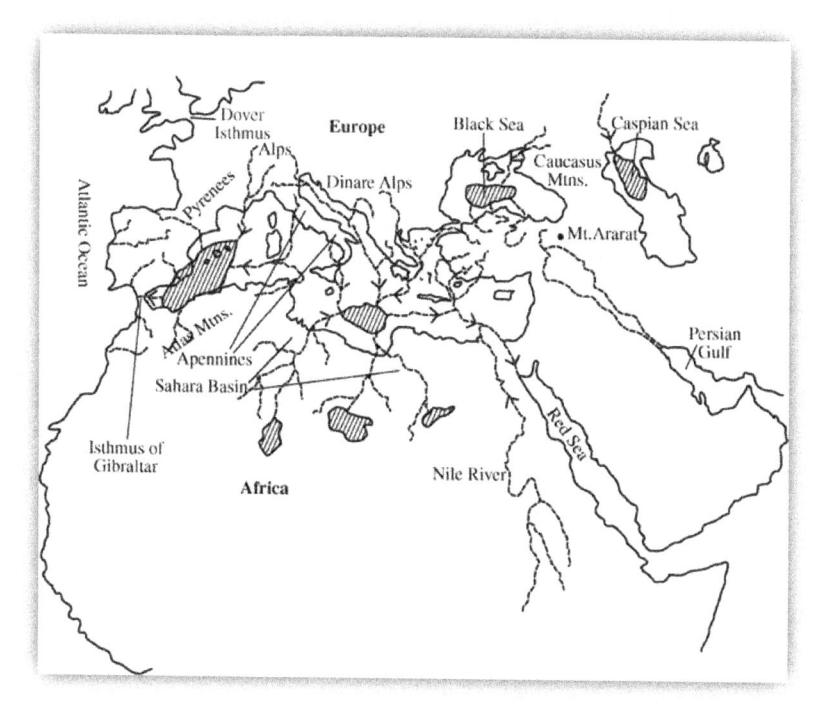

Revisiting Armageddon: Asteroids in the Gulf of Mexico

***Possible Configuration of Mediterranean Plains Region
Prior to 7,500 YBP***

As the land surface assumed recognizable shapes and elevations, there were still frequent collisions of material that were dropping out of orbit as they collided with each other and eventually came crashing onto Earth's surface. Those that landed where today there are oceans are just about impossible to detect today, but some of them landed on dry land, either singly or in huge groups. Evidence of some of the larger ones can be seen today.

One of these falling materials is now Greenland. This island is now really a large crater. Its perimeter is recognizable as a shoreline today, but the center is a deep pit, now filled with ice and thus resembling an island, as we would normally picture one.

If the materials that struck Earth and formed Greenland, had landed in the South Pacific Ocean instead of the Arctic Ocean, it would have looked like the largest atoll on the planet. The debris from the impact would, as with Greenland, have dumped an abundance of material along the edges, forming a sea in the middle of the South Pacific.

A second example is the Mediterranean Basin area. High land formations to the north and south sides of the central wave area helped to contain the water and slowly turn it from ENE to east, on a path toward the east end of the Mediterranean Basin. The Atlas Mountains of the African shoulder in Morocco and Algeria were struck hard and had much of their volume spread out into Algeria and Tunisia. These were largely denuded, and their mountain tops lopped off, thereby reducing them in size.

Here in the western part of the basin, there is an interesting effect that shows the flow of the sea as it comes in from the southwest, flowed northeast at first as you would expect, before the mountain ranges on the north and south sides turned the flow more easterly.

Tunisia has a northeast facing shoulder unlike most other major land masses in the western Mediterranean Basin. The waters abutting Tunisia are shallow, and slowly deepen to the northeast. The flood left suspended material here in its wake as it sped over Tunisia and into the basin.

The Pyrenees Mountains of northern Spain and southern France helped to turn the main force eastward, as did the Alps of northern Italy, and the Dinare Alps east of the Adriatic Sea. Surely some of the water entered southern France, thus striking it from the

Revisiting Armageddon: Asteroids in the Gulf of Mexico

south as well as the northwest area we call the Cherbourg Peninsula.

From the Canary Islands all the way east to Turkey, west facing shorelines were eroded by the force of the water, leaving rocky, sometimes steep shorelines. These shores can be found in Morocco, Spain, Portugal, Minorca, Majorca, Sardinia, Corsica, Italy, Sicily, the eastern shore of the Adriatic Sea, Malta, many of the islands in the Aegean Sea, and the west facing shoreline of Turkey.

West Coast of Italy—Typical View

In Monte Carlo, Monaco, one structure near the Hotel de Paris has been built out over the water beyond the shoreline. Land here is scarce and priceless, and that must be why this has happened.

Any sections of these shores which ended up in a protected harbor configuration today, have sandy, shallow bottoms, caused by currents moving laterally along the coast, plus tidal actions, which dropped suspended materials in such coves when the water slowed down upon entering them.

In contrast to these western shores, east facing shores are just the opposite. In general, land extends slowly into the sea, and shallow waters lie offshore, not only off Tunisia as described above, but off the eastern shores of Italy, Malta, and most of the islands in the Mediterranean Sea.

The area of the Mediterranean Sea south of Monaco and west of Italy is called the Ligurian Sea. When the main force of the flood struck here, it raced across Northern Italy and formed the Po Valley from the material the water had been carrying in suspension. This is the low, fertile land north of the mountain range that runs down the Italian Peninsula, and south of the Italian Alps. It dumped great amounts of silt and other material and covered this area.

The western shoreline of Italy from Monaco south has a steep, rocky shoreline except for a few indented bays which formed natural harbors much later than the flood. Indeed, in order to build large protected harbors for commercial ships and pleasure boats, it was necessary to erect offshore rock barrier reefs running parallel to the coast, to create protected waters for boats and ships.

The eastern shores of Italy are quite different from the

Revisiting Armageddon: Asteroids in the Gulf of Mexico

western shores. Absent, for the most part, are the rocky cliffs. Sandy shores extending out into the sea are the rule of thumb. When the flood roared over Italy, upon entering the Adriatic Sea, it slowed and then dumped much of the material it was carrying into the Adriatic Sea. This is why a typical depth in the central Adriatic Sea may run 1,000 feet, and off the western shore by the same distance, the depth averages about 10,000 feet. The exact latitude I checked was that of the Naples area.

The flood wiped out almost all of the people—all civilizations in its path. Those thousands—and maybe millions—of helpless souls never knew what hit them, literally. They certainly felt violent earthquakes and resultant fires immediately preceding the advance of the water, and then looked up to see a mountain of water, miles high come crashing down upon them, not only destroying all the humans, animals, and birds in its path, but also reducing all traces of their civilization to mud, gravel, and sand.

Anything man-made was utterly and completely destroyed, as far as I know. Items that could deteriorate over time, such as wooden objects, pieces of buildings, boats, and other structures would soon rot away completely leaving no trace.

The advance of the flood waters from west to east slowed and stopped in eastern Turkey and Armenia. This occurred when the western end of the 3,000 mile long wave finally got through the Mediterranean Basin and reached the eastern end of Turkey.

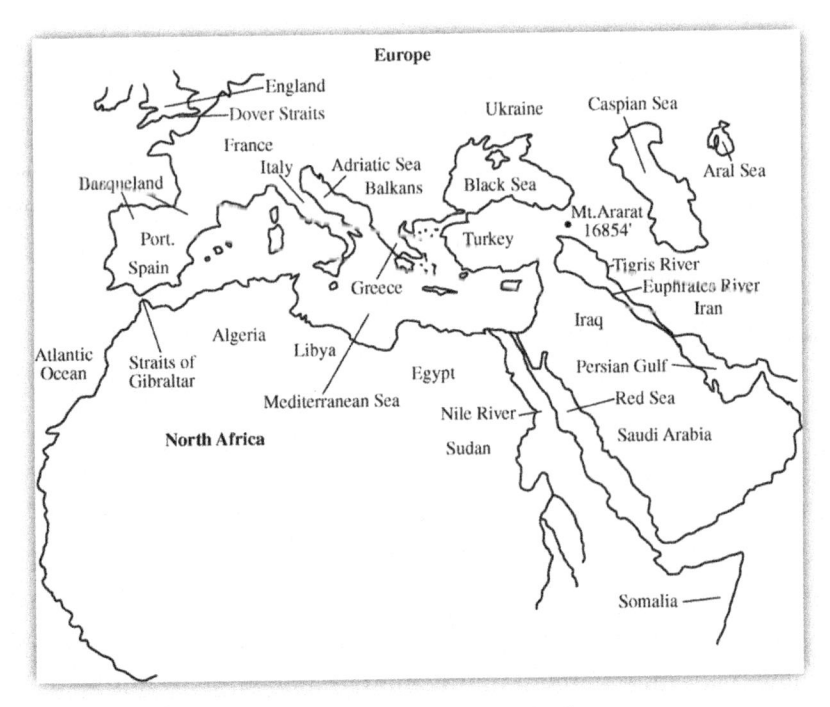

Mediterranean Region Today

Please note that the distance from Gibraltar to eastern Turkey and Armenia is about 2,000 miles. There was now no more water behind the great wave to push it along any further eastward. The water in this area would then have reached its maximum depth. We know it had to have been at least four miles deep to cover 16,945 foot Mt. Ararat, as described in the Holy Bible and numerous other sources and stories.

Just how much deeper than four miles the water reached is anyone's guess. When I think of the huge amounts of sea water that spread out from Turkey in all directions, it leads me to think the maximum depth was far greater than four miles.

If any traces of civilization which preceded the flood still

Revisiting Armageddon: Asteroids in the Gulf of Mexico

survive, they will most likely be found buried deep in the central sea area east of Tunisia, and at the bottom of the Adriatic, Black and Caspian Seas. It is highly unlikely anyone will ever recover any remnants from civilizations that preceded the date of the flood, or about 7,500 years ago, but one never knows. Maybe someday. Just maybe.

Ray Covill

CHAPTER TWENTY-SIX

Southern Flank of the Flood

The area running from the Straits of Gibraltar south along the African shoulder for hundreds of miles was impacted severely by the wave coming out of the Atlantic Ocean. The mountainous areas we now call the islands of Minorca, Majorca, Sardinia, Corsica, Capri, Sicily, Malta, and others, were all cut down in size and, to a lesser degree, so were Crete, Cyprus, and the Aegean Islands.

The Atlas Mountains run along a southwest to northeast line through Morocco. They are not at this time a major mountain area of grand proportions such as the Alps. Indeed, the high point of Africa is not in the Atlas Mountains at all. It is Mt. Kilimanjaro in east Africa, which is a volcanic peak topping out at 19,340 feet.

The Atlas Mountains stood right in the path of the great flood as it raced to the northeast and east. As the frontal edge of the wave approached Africa, it slowed down a bit due to friction from the seabed in the shallow offshore waters. This would have caused the sea to rise even higher as it crashed into the Atlas Mountains, ripping up everything loose in its path and tearing away at the higher, more vulnerable peaks and flanks and plucking away at the eastern flanks, the same way that a glacier does.

Revisiting Armageddon: Asteroids in the Gulf of Mexico

The seas crashed into and over the coastline of North Africa and then inland into the mountains. The Atlas range did not stop the flood. Rather, the water just went over the mountains and tore away much of the loose rock and earth, taking them hundreds of miles east into Algeria and Libya.

Today the landscape in the northwestern reaches of Africa is littered with loose stones, large rocks and escarpments. The force of the water reduced the size of hills, eroded the landscape, and cut channels into the sand and gravel soil, creating escarpments in a haphazard assortment in many parts of Algeria and eastern Morocco.

The Sahara Desert area, from the Atlas Mountains eastward all the way to the Red Sea, was once a fertile land with a warm climate and a viable population, probably agrarian in nature. When the flood struck, all the countries of North Africa, from Morocco on the Atlantic to Egypt in the east, were buried in sea water and sand from the Atlantic when a section of the ocean charged into the Mediterranean Basin and into North Africa. As the water raced eastward it scoured the land in some places and buried many depressions in its path—including rivers.

That's correct: rivers in the Sahara.

NASA satellites have discovered complex river systems in parts of Libya and Sudan. They seem to run northerly and may have connected in places to become a system of rivers across North Africa, draining into the Mediterranean area, which some then drained westerly to the Atlantic Ocean, and some easterly to the Red Sea and the Indian Ocean.

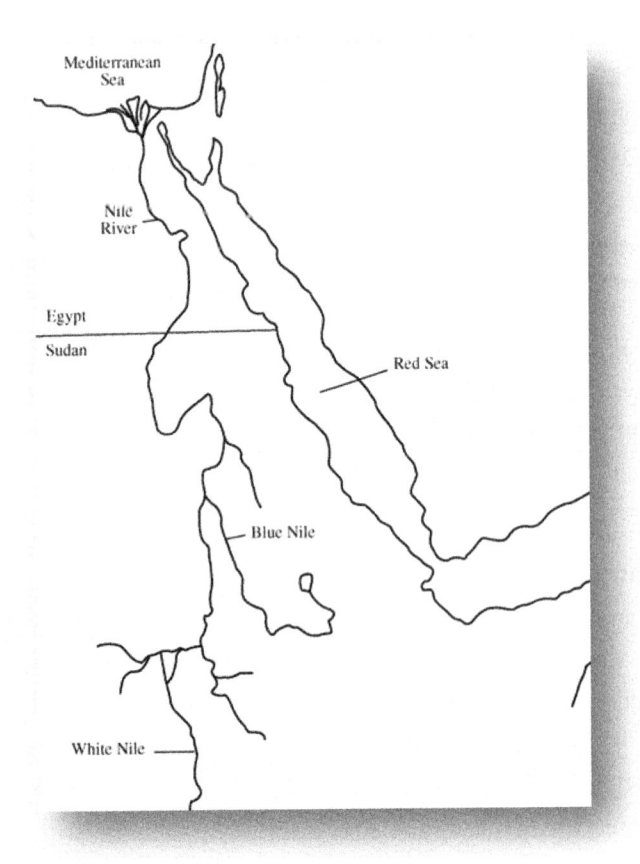

Nile River Area—Egypt/Sudan

But assuming this system did in fact drain North Africa, it probably drained northerly and then eastward to meet up with other rivers in the lower Mediterranean Basin, north of the Nile Delta and on to the Red Sea, where it drained into the Indian Ocean. If it ran westward it would have encountered the Atlas Mountains.

It was NASA satellites which detected an underground system of rivers that partially covered some of this area. The water is as close as one inch to the surface in many places, making it detectable by satellite imagery. In some lower areas the Sahara

Revisiting Armageddon: Asteroids in the Gulf of Mexico

rivers still reach through the surface of the sand to create bodies of water we call oases. These oases have enabled human travelers to cross over this land and live in this barren desert. The North African buried river system appears to have its headwaters in the Sudan area, although some of it may have come from the eastern flanks of the Atlas Mountains.

Some rivers emanate from underground water sources, but most major rivers have above ground sources. These always start out from either a lake or some other major body of water, including an inland sea. Rivers may also have their headwaters in mountain ranges.

The mountains attract rainfall and/or snowfall and the resultant abundance of localized water follows the lowest point on a downhill journey to the ocean, or occasionally a large inland sea such as the Caspian, beyond which there is no place further to run off downhill.

The river in the Sahara Desert has no name, even though the people who inhabited the area thousands of years ago probably named it at one time. This name, however, has been lost through time. So, since I believe its headwaters were somewhere around Sudan or the Atlas Mountains, I'll refer to it as the Sahara River, for lack of a better name.

The Sahara River was not a single strand like the Nile; it had a multi-branch system of headwater sources similar to that of the Amazon River in South America. This river gave dwellers of the area a reliable water source.

The Nile River originates in several lakes in East Africa and runs northward to the Mediterranean Sea, while the Tigris and Euphrates Rivers start out in the mountains of western Turkey and run somewhat parallel courses southerly, then meet and go to the Persian Gulf and the Indian Ocean. These are all "single strand" rivers because there are no adjacent areas of water systems or lakes that could feed into them to form a network of tributaries.

Chapter Twenty-Seven

Wadi Kufra

In 1978 and into the 1980s, NASA started experimenting with ground penetrating radar capabilities in its satellites. They detected water under the top layers of sand in parts of the Sahara Desert.

On October 4, 1994, using the Spaceborne C/X-band Synthetic Aperture (S/R-C/X-SAR) imaging radar on the sixtieth orbit aboard the Space Shuttle Endeavor a system of old, now inactive stream valleys was found—which, I believe, carried running water northward and eastward across the Sahara.

Wadi Kufra—Near Kufra Oasis, Southeast of Libya

This region is now hyper arid, and receives only a few millimeters of rainfall per year. The valleys are now dry "wadis" or channels, buried by windblown sand. The west branch of the system, known as "Wadi Kufra" (the dark channel along the left side of the image), was recognized and much of its course outlined.

The broader east branch of Wadi Kufra, running from the

Revisiting Armageddon: Asteroids in the Gulf of Mexico

upper center to the right edge of the image, was unknown until the SIR-C imaging radar instrument was able to determine its size (about three miles by sixty miles). The two branches converge at the Kufra Oasis, the grouping of circular fields at the top of the image. The SIR-C image shows river channels cut into the bedrock. This would have required a huge flow of fast moving water carrying rocks and boulders to chop away and cut into the bedrock, defining the direction of the water.

The area shown is about seventy-four miles by forty-eight miles with the upper left serving as North. The colors shown in this image were based on the following radar channels: red represents the L-band (horizontally transmitted and received); green represents the average of the C and L band; and blue represents the C band (both horizontally transmitted and received).

Another NASA discovery of hidden rivers is discussed in the same NASA bulletin. This one is in an uninhabited area called the Safsaf Oasis, also in North Africa, in southern Egypt, not far from the Sudan border. This discovery was also produced from data obtained from the L-band and C-band radars that are part of the Spaceborne Imaging Radar C/X-band Synthetic Aperture Radar (SIR-C/X-SAR) onboard space shuttle Endeavor on April 9, 1994, a few months before the Wadi-Kufra discovery. The image is located at 22° North Latitude and 29° East Longitude. The dark blue wavy lines are various size channels and streams that cover an old, broad river valley.

The radar signals easily penetrated up to an average of five inches of surface cover of loose, windblown sand. Just about everything visible on these two radar composite images cannot be seen, either when standing on the ground or when viewing photographs or satellite images such as those from the United

States' Landsat satellite. The radar interface likely shows the shallow underlying surfaces of river gravel or bedrock which are generally covered by only a few inches of windblown sand.

Safsaf Oasis—Southern Egypt

Who would have thought—even as recently as thirty-five years ago—that scientists could detect hidden river systems below the surface of the land? It is simply amazing. Perhaps if NASA should someday make other detailed surveys of the North African surface from Morocco east through Algeria, Libya, and Egypt, all the way to the Red Sea, maybe they would be able to map the Sahara River system in great detail enabling geologists to recreate the scene as it would have appeared before the Great Flood from the Gulf of Mexico struck some 7,500 years ago.

Ray Covill

Chapter Twenty-Eight

Kinks in the Nile

NASA's Spaceborne Imaging Radar efforts in 1994 revealed another exciting discovery in North Africa. The Nile River runs a general course to the North until it finally empties into the Mediterranean Sea at Alexandria. The course takes it through mostly sandy desert. In northern Sudan the Nile very suddenly abandons its northerly flow and swerves off to the southwest. Then it returns to the north, moves easterly, and finally resumes its northerly course to the Mediterranean Sea. It is a long looping "S" detour of around 200 miles.

NASA discovered an ancient river channel buried under shallow layers of sand revealing an original course direct to the north. How could this change in course have occurred in a river passing north through the Sahara Desert, the largest and driest desert on our planet? There must be a reason for the river to make this change in course.

The Great Flood's southern flank overran Egypt and Sudan, an area bordering the southern limits of major flooding activity. In Sudan and southern Egypt the slower moving water carrying huge amounts of sand and silt came to a near stop and dumped a large amount of material into and around the "old course" of the Nile where it had once flowed northerly through Sudan.

Not only did the flood waters bury the river in this area, but the front wall of the flood dumped even more land material to the east between the river and the Red Sea along the frontal boundary of the tongues of water as they finally slowed and stopped. This action is similar to a terminal moraine formed by a glacier which marks the place of maximum advancement. A terminal moraine occurs when ice and earth are being pushed, however slowly, by a glacier until it stops and melts, leaving a jumbled strip of landscape at the frontal boundary of the glacier.

A good example of this kind of activity can be seen in southwestern Cape Cod, Massachusetts through north and west Falmouth, all through the Elizabeth Islands, and further to the southwest to a point called Brown's Ledge, underwater. As big chunks of ice melted they left "kettle holes" where the ice once was located. The area of Massachusetts described above is full of small hills and kettle holes. In addition to this creation of terminal moraines, earth and rocks that had been suspended throughout the surface of the glacier, dumped material in place as the glacier temporarily stopped and melted, creating rock islands and shallower water in the case of Buzzards Bay which abutted the north side of the terminal moraine. This type of rocky terrain is called a "recessional moraine" and is the reason Buzzards Bay is shallower than Vineyard Sound to the south of the terminal moraine.

Once this action in the African desert ended, the surplus water drained off in all directions leaving behind a brand new landscape with radically different topography. Meanwhile, the Nile's waters were still flowing north through southern Sudan and suddenly found no riverbed to follow. So what occurred next is very plain to visualize. As the flowing river water was blocked from continuing north, it began to pile up at nature's new dam, creating a huge backup lake stretching hundreds of miles back into southern Sudan.

Soon the water spilled out of its banks flooding the lowest land areas to the immediate west and east of the river, similar to the way the Nile floods Egypt every spring, or comparable to the Mississippi River when heavy rains occasionally raise the water level high enough for the river to run over the banks that normally contain it.

In Sudan, the lowest adjacent land areas were apparently to the southwest of the dammed up river, and the water, slowly at first and then developing a more definitive course, started a run to the southwest. When further movement in this direction found lower land to the north, the course changed to this new direction. Similarly, when further northerly passage met with land of a higher altitude the water stopped, backed up again, and eventually found a new "lowest exit" to the east and started to trickle off in that direction, consolidating adjacent side stream trickles until it had established a definite easterly course.

Then, when the eastern flow ran into higher land deposits it built up and deepened again until it found the old northern course of the Nile that carried water to the Mediterranean Sea before the Gulf Event occurred. As more water, at a faster pace, began to entrench the new "S" detour, the Nile firmly established this new turn and runs the course we know today.

A second, but much smaller "kink" in the Nile occurs in northern Egypt. It is found about 400 miles south of the Nile's terminus at the Mediterranean Sea. It is in a comparatively small, thirty-mile bend to the east, then a similar distance north, and another back to the west to rejoin the river's "former" path It would be interesting to see if NASA can detect the buried original straight northerly course of the river.

The flood dumped enough material at this point to block the river and force it to reroute itself as described above. Bedrock must lie well down at this point since it existed before the flood, and there may be tombs built at both "kinks" far below the deepest ones known to the Egyptians of today. They could be as deep as the level of the "old river" or close to it. Below this point, an excavation would encounter groundwater that had been trapped since the flood.

The volumes of water near the terminus of the flood in the Egypt/Sudan area, the speed of the water, and the amount of sand and silt it was carrying as it slowed to a halt, as explained above, all left their mark. There is no reason these factors should be uniform. The vagaries of nature shot many arms or lobes of the flood in varied directions with an array of different water volumes, speeds, and amounts of sand and gravel being carried at any one time and place. The movement is all based on the topography of the land the water sped over, as well as the proximity of any one lobe to the central thrust area of the flood.

In any case, huge amounts of sand were picked up by the flood and even larger amounts of sand and gravel were ground down into fine sand due to the friction of its being dragged along for many hundreds of miles. As the flood slowed and stopped, it left sand deposits all over North Africa from Algeria to Egypt. It does not appear to me that this side wash, south of the main thrust of the floodwater through the Mediterranean Basin, reached the

Red Sea in any significant amounts.

The land area bordering the west side of the Red Sea appears to contain low rolling hills compared to the area to the west of the Nile. Perhaps this "barrier" finally stopped the now slow moving easterly flow of the flood in Egypt and Sudan, or else it is the delineation of the front lobes of the flood, which showed where the "terminal moraine" type deposits ended up.

The nature of the above described Nile detours indicate that the floodwaters were petering out in this general are of the Nile, from the central Sudan north through Egypt to the Mediterranean Basin. But the flood left all this evidence behind for us to ponder today.

CHAPTER TWENTY-NINE

Egypt's Great Sphinx

As the flood roared through the Mediterranean Basin, it destroyed man-made structures and almost all living things including people, animals, birds, vegetation, and forests. Most were destroyed upon impact while others were smashed and buried in the mud and sand in the basin where anything living soon rotted away.

Amazingly, there were two large man-made structures that survived. One was the Great Sphinx in northern Egypt, and the other I will discuss later on.

The Great Sphinx

The Sphinx was carved out of a local outcrop of limestone bedrock that protruded above the desert floor. It has the body of a lion and the head of a pharaoh and measures 240 feet long and 66 feet high. The side view of the Sphinx has become so familiar to us that we accept it as an art object made by sculptors in the days of the pharaohs.

The surface condition of the Sphinx gives a clue as to the relative ages of its parts. The head shows damage to the face and mantle, which could be from vandals before and after the time of Christ. The body shows severe erosion, both laterally and vertically. This is the original surface of the body and there is nothing to indicate it was ever repaired. The front legs were "re-plated" with limestone slabs by the Egyptians and the seams between the slabs are readily discernible and show little wear over the last 4,000 years or so.

The longer I look at the side view of the Sphinx, the more I am convinced that there is something very wrong here. The proportions of the front legs to the body are correct, the height and width of the body to the hunched up rear legs are all in order, but the head doesn't belong to the body. It should be ten times larger to be "in sync" with the rest of the form.

I find it difficult to believe that a sculptor and his team, probably the most talented group available at the time of construction would carve a well-proportioned lion, but with a head so small it looks like an afterthought.

When I was very young, my parents erected a Christmas tree in our living room every year from mid-December to January 1. The height ranged around six feet and it was crowned with a silvery star about nine to ten inches in diameter. Why that size?

Revisiting Armageddon: Asteroids in the Gulf of Mexico

Because it was appropriate—it just looked right. They never put a star on top the size of a belt buckle because it just would not look right.

When the Sphinx was first carved, the head was that of a lion. The entire object was a lion. So what happened to the lion's head? The answer is conjecture on my part, but it is the only plausible explanation I can think of.

How the Sphinx may have looked at original construction 10,000-20,000 YBP

The Great Sphinx was built between 10,000 and 20,000 years ago, and maybe even longer. There is no way to know exactly when the Sphinx was built, but I predict no more recent than 10,000 years, and no longer than 20,000 years ago. I simply can't imagine it surviving normal, slow processes of erosion and earthquakes much longer than that; however, it is always a possibility that it could be as much as 30,000 years old.

When the flood rolled over Egypt 7,500 years ago, the force of the water and debris it was carrying slammed into the Sphinx and broke off most of the exposed head. This was the part that stood out and would, therefore, have been the part most susceptible to damage from the flood waters.

The front legs of the statue were apparently damaged to some degree also. The body shows parallel, horizontal striations, or erosion surfaces along with a few vertical run-off indentations, all of which could only have been caused by huge amounts of water racing over and around the statue. This damage could not have been caused by rain over the centuries, most obviously because rain almost never occurs in the area. Neither could it have been caused by repeated sandstorms. That would have caused more uniform erosion and rounding all over the statue. The newly repaired front legs show no such evidence of erosion by water. This is because the legs were repaired long after the flood occurred.

The head that we see on the Sphinx today shows some deterioration, but nothing like the erosion damage to the body. To me, this indicates that the head, as it appears today, is much newer than the rest of the sculpture.

When the easterly motion of the flood stopped, a period of perhaps a few months of receding drainage followed. This flow caused many additional horizontal, layered, erosion striations on the sides of the lion, as well as the vertical run-off indentations.

All this leads to the obvious conclusion that the flood caused the damage to the original head, as well as on the sides of the lion's body.

Revisiting Armageddon: Asteroids in the Gulf of Mexico

After severe flood damage 7,500 YBP, Pharaoh Tuthmosis, 3425 to 3417 YBP, reshaped what remained of the head into a likeness of King Khafre

One thing that was true thousands of years ago and is still true today is the recognition of the lion as the king of beasts. The sculptors did not build a giant monument in the likeness of an elephant, hippopotamus, water buffalo or rhinoceros. The lion was king then, as now. Today, we regard the lion as the king of beasts. The people of the Sphinx era had come to the same conclusion and so they sculpted the Sphinx into the form of a lion.

The people who built the Sphinx obviously had detailed knowledge of the larger animals of central Africa. They had probably ventured up the Nile by boat and then explored all over East and Central Africa.

The next question is how and why did a pharaoh's head get sculpted on the Sphinx? The answer may be as follows: The ugly, broken lion's head remained untouched after the flood for 3,000 years or so. Sometime during that interval, people migrated into the Egyptian area where the prior civilization had been utterly wiped out by the flood.

The current face of the Sphinx is believed to represent

King Khafre who reigned for sixty-six years, starting in 2558 BC. When the early pharaohs came into power, starting around 4,500 years ago, Pharaoh Khafre, who must have been a very vain person, got the idea to recut his image onto the edifice. The resulting head was of course much smaller than the lion's head because there was a lot less rock remaining to work with. That is why the Sphinx's head is so out of proportion compared to what one might expect if the entire Sphinx was carved at the same time.

Lion Statue

Perhaps that pharaoh reasoned that a small head was better than no head at all and certainly better than the broken, remains of the lion's head that, after the flood, was left as an unrecognizable lump.

Try and recall the looks and size of the "Old Man of the Mountains" in New Hampshire prior to its recent collapse. If someone wanted to re-carve a new head using whatever solid rock remains today, that head would be tiny compared to the original.

A similar dilemma confronted the above mentioned

pharaoh but he chose to proceed with what was left of the Lion's head and so put his head in its place, however disproportionably small it would be. And so it was done. About a thousand years after Khafre's death, Pharaoh Tuthmosis IV (born 1425 BC) restored the Sphinx's legs, and he also added a royal statue at its chest.

In 1987, a team of Japanese investigators, using earth-penetrating radar and electro-magnetic equipment found hollow cavities and tunnels deep under the body of the Sphinx. No further examination or excavation has been allowed by Egyptian authorities.

It makes one wonder what is, or was, inside the Sphinx. Could it have been another tomb such as those found inside the much more recent pyramids and in the Valley of the Kings? If so, who may have been buried there? Was such a tomb looted like most of the others, or did it escape the grave robbers, and survive until now, intact? Or could the cavity and tunnels contain precious historical relics placed there more recently than the pyramid builders? Or could it contain artifacts from an ancient civilization that pre-dated the Great Flood? No one will ever know until the Egyptian government allows its antiquities researchers to excavate, find the entrance, and take a look inside. I hope I live long enough to see that happen.

Its importance could dwarf that of the discovery of King Tutankhamen's tomb in the early 1920s or the finding of the Dead Sea scrolls in the 1940s.

Chapter Thirty

Basque Land

When Americans hear the word "Basque," the first thing that crosses their minds is usually the game of jai-alai, a game invented by the Basques. The game has taken hold in parts of the United States and is played in large, lavish gambling casinos by Basques. However, there is much more to the Basque people and their history than a casino game.

The origins of the Basque people are unknown. They live in northern Spain and extreme southern France, and were the largest group of people in the Mediterranean Basin area to survive the flood. That is to say, there were enough survivors to enable this group of people to continue on, reproduce, and flourish, which they have done to this day.

They now live mainly in the Pyrenees Mountains of Northern Spain, and the existence of these mountains may have been their savior. If that area had been a flat, open land, none of them would have survived. The fact that they did survive is rather miraculous. It was most certainly helped by the fact that they were located north of the main thrust of the flood, and had partial protection from the mountain range where they lived. It is highly likely that a huge majority of these people perished, but the remainder succeeded in carrying on their race.

The Basque people have preserved their national identity and successfully resisted foreign invasions, influences, and assimilation. Today they number around two-and-a-half-million people, and occupy a small, seven-province area of northern Spain and extreme southwestern France.

The Basque language is ancient and has survived with little change to this day. It is called "Euskera" and is the oldest language still in use today. It bears no relationship to any other language on Earth, current or ancient, and proud as they are of their language, the citizens refer to themselves not as Basques, but as "Euskera speakers." Still, the first book written in Euskera did not appear until 1545.

A Spanish historian in the 1500s claimed that the Iberian Peninsula was inhabited by descendants of Tubal, Noah's grandson, who went to Iberia thirty-five years after the flood had subsided. Many legends abound.

One stated that Tubal, or Aitor in the Basque language, was one of the few men who survived the flood who was not in Noah's ark. Tubal is recognized by some as the father of all Basques.

Linguists who have studied the language find that the grammar of Euskera has resisted change, unlike many other languages, so that the modern version of Euskera is not unlike its ancient form. It has complex verbs, twelve cases, few abstractions, and no prepositions or articles.

No one has ever found a language even remotely related to Euskera. It is a strange tongue which pre-dates the Indo-European family of languages. The Basque language may be a pre-Bronze Age tongue, and is apparently the oldest European language. If this is so, Basques could then be the oldest European culture.

Basques have developed and preserved not only their own unique language, but also their own literature, music, legal system, food, and traditions longer than any other nation in Europe. They are physically distinct, burly people, with many featuring a long straight nose, thick eyebrows, a strong chin, and long ear lobes. In past eras, when French and Spanish folk were typically small people, Basque men were bigger and had large chests, necks, and shoulders. This disparity still exists today.

These were also characteristics of Cro-Magnon people, and so Basques are often thought to be direct descendants of the Cro-Magnons who lived 40,000 years ago. These ancients are known to have lived in such other places as the Atlas Mountains of Morocco and the Canary Islands off Africa's western shoulder.

The fact that the Iberian peninsula and Morocco were once connected by a land bridge, or isthmus, lends further credence to the idea that the region was all one area inhabited by these people. I will discuss this in further detail later on in this story.

The Cro-Magnons and their Basque descendants also occupied the western half of what is now the Mediterranean Sea, but was then a land of rolling hills, lakes, and rivers.

Other physical evidence of an ancient and distinct group is that all their blood types are either A, B, or O. Over fifty percent of Basque people have type O blood. This is the highest concentration of type O in the world.

Most of the rest have type A blood, with B being rare. In

Revisiting Armageddon: Asteroids in the Gulf of Mexico

1937, Rhesus factor (Rh positive or negative) was discovered. Basques were found to have the highest rate of Rh negative blood of any people in the world. Those living in the Atlas Mountains in Morocco and the Canary Islands also have a high rate of Rh negative. This lends some credence to the Cro-Magnon theories.

We know nothing of the details of Basque life before the Great Flood, or after it for about 5,000 years or so. The Basque country is the oldest nation in Europe which has never actually been a country—Spain still claims all of the Basque Provinces as part of Spain. This has resulted in a "stand-off" relationship within the country for many years. The Basques do not covet any other parts of Spain or France aside from the Basque land on which they live; they've simply wanted to be left alone.

The fact that they have never been a warring nation bent upon conquering other neighboring countries, has been paramount in their search for independence. The fact that they still exist today is remarkable.

Over the centuries, the Basques have fiercely defended their land against invaders. Armies trying to invade have found great difficulties in negotiating streams, narrow paths, and steep terrain, only to find themselves attacked by Basques who suddenly appeared from nowhere, killing and harassing the invaders, and then quickly disappearing into the hilly terrain, guerilla style.

Even the Romans gave up the idea of conquering the Basques, and instead became trading partners with them. Their reputation as fighters has made them highly prized as mercenaries for several thousand years.

In 1936, Spain's Franco was allied with Adolf Hitler, who helped General Franco in Spain's civil war. The Germans provided troops, planes, and weapons to General Franco, a poor military tactician. When Franco and his troops failed to take Madrid, the Germans stepped in to take command of the war and attacked the rebel Basques, whose entire war machine consisted of bolt-action rifles and pistols.

On April 6, 1937, a swarm of Heinkels, Junkers, Derniers, and Messerschmitts made a low-level attack on the Basque town of Guernica and leveled it with splinter and incendiary bombs and machine guns. Twenty percent of the population died. Later, another fourteen towns were bombed.

The world was horrified and outraged at the massacre of unarmed civilians, but little did they know that the Germans were only practicing in Spain—getting ready for World War II. Franco denied the attacks ever took place until in 1970 his government admitted to their war crimes for the first time.

In 1998 the German government finally apologized to the Basques. It took the Spanish government until 1999 to admit that Franco lied about Guernica. When WW II started, the Germans, French, and Spanish kept close watch over the passes along the Basque border. The Basques knew the footpaths, and could go up and down without being seen. They played an important role in World War II from May 1941 to July 1944 in smuggling 700 rescued American, British, and Canadian fliers whose planes were shot down over Western Europe, and then returning them to England. In addition to fliers, many political refugees and Jews were smuggled into Spain and to freedom.

About 1,700 Basque agents and the French underground were involved in the effort, which was called "operation comet," or to the French resistance, "la ligne" (the line).

Revisiting Armageddon: Asteroids in the Gulf of Mexico

Most of the fleeing flyers were smuggled at night to Brussels, Belgium dressed in the local garb, and then taken south by train to St. Jean-de-Luz on the Basque border. Female agents sometimes called "handlers" watched out for them during the trip. At Bayonne, or at St. Jean-de-Luz, the fliers exited the train station under the nose of the German inspectors and through the men's room which had a back door to the street. Here they were met, fed and rested in either of two local hotels. Then at night, Basque agents led them to a nearby farm, and then along small streams, paths and fields, and over a mountain. Travel was slow and tiring.

They moved in the dark to avoid the Germans who had ordered a blackout at night. Hours later, dressed as Basque peasants, they would be on the Spanish side of the border where passage was a bit easier and where the blackout did not apply. They then had to elude the Spanish civil guards en route to a farm where they could rest and eat.

They walked for miles on back roads and trails to the town of Renteria where they would take a train to Hermani and on to San Sebastian to the homes of other Basque agents.

The trip from St. Jean de-Luz to San Sebastian would today take twenty minutes by highway, but back then would take much longer depending on water levels in the rivers.

The final destination of the Basque section of la ligne was the British Consulate in Bilbao. Here, the British would get the airmen back to England via Lisbon, Portugal, or Gibraltar.

Basques participated in the D-Day invasion of France in June 1944, and by early 1945, Basque soldiers had cleaned out Germans who were in fortified positions along the southwestern French coast. If the Basques were ever to gain independence and form a new country of their own, it most likely would be named "Basqueland" since Europeans, and their ancestors had developed the habit of naming many of their countries with names ending in "land."

The Vikings had their Iceland, Greenland, Baffinland, New Foundland, and Vinland. Other more recent "lands" include Ireland, England, Scotland, New England, Rhode Island, Maryland, Holland, Netherlands, Shetland, Finland, Poland, and New Zealand, among others.

The only other name that sounds appropriate would be "Euskeraland" but that just doesn't sound right to me. Time will tell if they ever get the opportunity to gain independence.

Chapter Thirty-One

The Final Push

The distance from the Gulf of Mexico to Gibraltar is about 3,000 miles, and the distance from Gibraltar to the Caspian Sea is about 3,000 miles. When the Atlantic "slice" crashed into the Mediterranean Basin, the force of the water behind it kept pushing it eastward. When the western end finally reached Morocco, there was no more force behind it to keep the wave—which had become a flood at this point—moving eastward except its own momentum, which was a considerable factor.

As the flood passed through the Mediterranean Basin, friction with the land would also tend to slow down the westerly advance. Most of the water moved easterly with smaller amounts spilling off to the north and south, such as in the direction of the British Isles, the Iberian peninsula, France, Northern Italy, and the Balkans, as well as the central Sahara and Sudan to the south. But the bulk of the water was driven eastward.

As this easterly momentum kept driving over Turkey, there was soon only one place for the water to go: up. This deepened the water over Turkey and Armenia to a height of at least four miles, and probably much more. This created an ascending slope to the sea surface, which helped to slow down the average rate of easterly motion because the flow was now pushing uphill, meeting more and more resistance from the forces of gravity. And gravity tends to pull water down, not up. This is the major reason why the flood's easterly flow was brought to a halt.

Since the length of the Atlantic slice was longer than the Mediterranean Basin length, by a ratio of about five to three, all this water would not fit into its new home. The available space to hold it was quite inadequate and this caused great side spillage on the terminus.

When this happened, all the water over the Turkish and Armenian region, under the control of gravity, began to drain off in every direction—even back westerly into the Mediterranean Basin, through and above the Dardanelles Straits, and into the Aegean Sea.

Eventually, as the flow slowed, sea level in the Black Sea would have dropped, and soon water in the Aegean Sea began to flow back into the Black Sea until the Black Sea level was below that in the Aegean Sea. This "settling" procedure would have taken place a number of times until the sea level in the Aegean; and thus, the Mediterranean Sea, and sea level in the Black Sea, got themselves sorted out and stabilized.

This may explain why fresh water in the lower half of the straits still flows westerly today and salt water in the top half flows easterly. This is the only place I know of where activity of this "two-way flooding phenomenon" occurs.

Excess water in the Black Sea and Caspian Sea raced not only easterly and westerly, but north over the Crimean region and into Ukraine. The mud and silt that dropped out of the water north of the Black Sea created a thick layer of rich loam over much of Ukraine as the water slowed and stopped, and ultimately drained into existing northerly flowing rivers, scouring out their beds, and then carrying the water into the Baltic Sea and Arctic Ocean.

Some of the water filled existing lakes in the area south of Stalingrad and north of the Crimean region. The water could not drain off any further and so changed these lakes into salt water bodies of water, the form in which they remain to this day.

But the general flooding of Ukraine left rich soil for farming. Today this area is known as the "bread basket of Russia."

The Caspian Sea absorbed much of the easterly runoff from the flood's peak in eastern Turkey. It appears that the flood went farther east of the Caspian Sea. The Aral Sea, which lies east of the Caspian Sea, and Lake Baikal, which is east of the Aral Sea in central Asia, were also flooded. It is entirely possible that the flood raced easterly over this relatively flat land as far as the Mongolian desert.

When the Caspian Sea was full, excess water also ran south and southwest into Iran, covering the northern third of that country. Over the ensuing centuries there was very little rain water flowing into it. The water stagnated and slowly dried up leaving salt deposits over a huge area of Northern Iran. Today, that region is called "The Great Salt Desert," and now you know why.

Sea water entering the Caspian Sea also went north and northeast covering adjacent land in these areas adjoining the Caspian Sea. This would have made the entire area of the Caspian many times larger than it is today. Ensuing drain off and evaporation eventually reduced the Caspian Sea to the size it is today.

To the south and southwest of Turkey, the Tigris and

Euphrates Rivers, with their headwaters in the mountains of Turkey, ran southerly in almost parallel paths, until they joined south of Baghdad, and dumped into the Persian Gulf, and then into the Gulf of Arabia. The scouring process caused the channels to become incised and well defined in the process.

Much of the water draining southwest off Turkey, drained into Iraq as well as other Middle East regions, dumping out huge amounts of sand as well. Similarly, the Gulf of Aqaba as well as the Red Sea's northern end were also scoured out, but as the waters eventually slowed, most of the drainage ended. The water dropped out enough suspended material to partially block up the northern end of the Red Sea and the Gulf of Aqaba.

The final effect of the last of the seas flowing into these regions at slower speeds caused great erosion and the settling out of large amounts of sand, which blocked up the former headwaters of the Red Sea's northern end, and also made the Gulf of Aqaba relatively shallow.

It is interesting to note that areas in this region receiving fresh water runoff have been able to wash out the salt and become fresh water bodies.

To the south of Turkey lies the Dead Sea in Israel, so named because of its very high salt content. Very few things can survive in this environment. This sea, of course, was filled up by the flood. Excess water poured southwest into the Gulf of Aqaba and thence eventually into the Indian Ocean.

The fresh water streams dumping into the Dead Sea over the years have mostly been diverted for human uses, so that now no water drains out of the sea. As a result, as evaporation continues, the sea level drops and the salt concentration increases.

Revisiting Armageddon: Asteroids in the Gulf of Mexico

Before the end of this twenty-first century, we may see the Dead Sea dry up completely and resemble the Great Salt Desert of Iran, or the Bonneville Salt Flats in Utah, U.S.A.

It is important to note that all these current or past salt water covered areas—meaning lakes and a dried up desert—are scattered over a large area centered in eastern Turkey and Armenia. The reason for this is simply that this area was the center and high point attained by the flood waters when the easterly push finally ceased.

The runoff intensified as gravity pulled the water downhill in every direction. Because of the huge volume of water involved, the speed must have been great as the rough seas rushed along. As the speed of the water slowed, the suspended mud, silt, sand, and gravel were deposited along the water's path.

This would have continued for hundreds of miles in all directions, and did not abate until the seas over Turkey had drained off and some type of normalcy ensued. The resulting landscape had been eroded and cleared of vegetation, and even the steeper, mountainous areas of Turkey and Armenia had been worn down so that the land was left barren. It would have been many years before new vegetation and reforestation restored the area to something resembling its pre-flood appearance.

When the flood ceased its easterly motion, the Mediterranean Basin had become a sea. Most of the water stayed there, forming the Mediterranean, Aegean, Black, and Caspian Seas, rather than draining back into the Atlantic or Indian Oceans.

Chapter Thirty-Two

Caves

Two or three times over the years I have heard of cave discoveries in the Mediterranean area. These caves are near mountain summits and not on the lower sides or near the bases of these mountains. In most cases the caves are big enough for people to walk inside them and some of them even have interconnecting passageways.

The question asked by most historians and archeologists is, who made them, when, and why? The age of the caves still remains a mystery. They are barren with no trace of anything man-made inside them: no art work on the walls, no stone flooring, no evidence of furniture, no clues at all.

I believe the caves were dug, probably hurriedly, about 7,400 to 7,500 years ago, possibly by some pitiful survivors of the flood, but more likely by people who lived in Northern Europe, West Central Asia, or Northern Africa. Such neighbors of the then new Mediterranean Sea may have come to explore the new "ocean" and carved out the caves as resting or living spaces.

Once they had selected an area to settle in, they first would have dug these caves, high up on the adjacent hills or mountains. Their recent ancestors would have orally passed down, from generation to generation, stories of the flood and the horror and catastrophe it entailed. The caves served as a place of refuge into which people could run and hide with their families. They would have brought adequate food and water to escape an impending "second flood," should such a thing ever happen. They probably feared the coming of a second disaster.

These settlers would have theorized, or been told by their elders, that the first warning would be major earthquakes, soon followed by fires and then the flood itself.

The only locations I have heard of where caves of this type exist are on the summits of hills and mountains on the Canary Islands, Malta, western Turkey and the Dead Sea region of Israel.

The fabled Dead Sea Scrolls were discovered in caves well above sea level. It is possible that about 2,000 years ago the scrolls were carried by boat to a remote shore of the sea, unloaded and placed in a nearby cave system. At the time, such a cave would be a good hiding place.

As the Dead Sea level dropped over the next 2,000 years, the caves became more and more inaccessible. Then, about sixty years ago, some climbers decided to explore the caves they could see from the sea, so they scaled the cliffs and the famous discovery was made.

The reason I feel I am correct about these caves is that no better explanation has ever come forth that I am aware of, and secondly, my explanation is very plausible. If someone can come up with a better explanation, I would be very interested to hear about it. Maybe someday we will learn more on the subject.

Chapter Thirty-Three

Noah

Some of the oldest stories in the world rank as some of the most fascinating, in part due to their age, and in part because they are based to some degree on fact.

Some have been handed down orally over many generations and never been put into print, as described in Chapter One, while others have been put into some sort of tangible writing, whether on clay tablets, woven fabrics, or stone walls, in ancient languages, hieroglyphics, or pictures.

The story of Noah and his ark (barge would actually be more accurate), is well known to most people. Parts of it, I feel, are almost impossible to believe, but there are other parts that can be explained.

The story of Noah appears in the book of Genesis in the King James Version of the Bible, and it also appears in the books of other faiths. The story of a great flood is common in several cultures throughout the world, though there are many variations in the story. Spanish explorers venturing to the western hemisphere found that even American Indians talked about a great flood legend.

I believe the story of Noah, his ark, and the great flood,

Revisiting Armageddon: Asteroids in the Gulf of Mexico

took place 7,500 years ago during the same flood described in this book. In the story of Noah, his barge ran loose on the seas for almost forty days. When the seas calmed down, the barge ran aground high up on Mt. Ararat in Turkey, the summit of which is 16,946 feet above sea level.

Some centuries later, the barge was apparently damaged in an earthquake which broke it into two pieces, one of which slid down the mountain one to two thousand feet where it jammed up against some rocks overlooking Ahora Gorge. And there it rests today.

For hundreds of years, travelers have told stories about seeing a ship high up on Mt. Ararat. Some of the people who lived in the eastern region of Turkey claimed to have climbed the mountain to get an up close look at the ark.

Since then, many organized expeditions of explorers from all over the world have climbed the mountain in search of the ark. Some claim to have found what looks like the end of a ship sticking out of the perennial ice on the northwest side of the mountain at Ahora Gorge. And yet, so far no one has come up with any definitive proof of the ark's existence, such as close up photographs or pieces of wood that could be carbon-14 test dated. The Turkish authorities have not allowed any such pieces of the ark to be removed from the country.

It has been only in recent years, due to a climate warming period, that the ice on the ark has melted back enough so that the end of the ship can be seen protruding through the ice and snow. The ice has preserved the structure in remarkable condition over the years.

Some explorers have found pieces of wood in varying

sizes that appear to have been tooled and shaped by humans. These artifacts were found in the outwash plains in front of glacial streams at lower elevations on the mountain below Ahora Gorge, but their location today is unknown. The possibility exists that they may be broken off pieces of the ark itself.

The Bible says "all the animals in the world were gathered," which reminds me of a Christmas morning many years ago, when I was about three or four years old. Santa brought me a foot-long ark with two lions, two tigers, two elephants, two giraffes, two gorillas, and two rhinos. Mom explained the story of Noah to me, but it probably went right over my head. My ark had four wheels and a string attached to the bow, and I towed that thing all over the house.

The biblical story of Noah as it appears in Genesis says the animals entered the ark "two by two." In looking back at the story analytically and not taking it all on faith alone, I don't see how Noah could have gathered up two of every species of all the animals on Earth. He would have had to collect various species from all over the planet and in the given time, this seems impossible. He would have had to go to Africa for elephants and other wild animals, to the Americas for buffalos, llamas, polar bears, and brown bears, Australia for kangaroos, New Guinea for white rhinos, India for tigers, Antarctica for penguins, China for pandas, and Kamchatka for white tigers.

And what about birds? There are now a little over 9,000 species still alive on Earth. In Noah's time there could have been 10,000 or more. A pair of every species? No, that would have been utterly impossible. Noah never even heard of 99% of the animals listed above nor 99% of the bird species. Over the years, the "two-by-two" statement has been interpreted to mean side-by-side—one

Revisiting Armageddon: Asteroids in the Gulf of Mexico

male and one female—of each species, as in a mating pair, coming in two abreast. But this can't be true either.

If the saying "two-by-two" did not mean one pair coming into the barge together, what did it mean? The answer may surprise you. It is really quite obvious, but we fail to recognize simplicity when we read too much into the story.

The animals came into the barge "two-by-two" simply because the doorway wasn't wide enough to allow them to enter three-by-three or four-by-four. When you stop and think about it, a hatchway about six feet wide would have been plenty wide enough to get the animals aboard, either singly or some coming side-by-side.

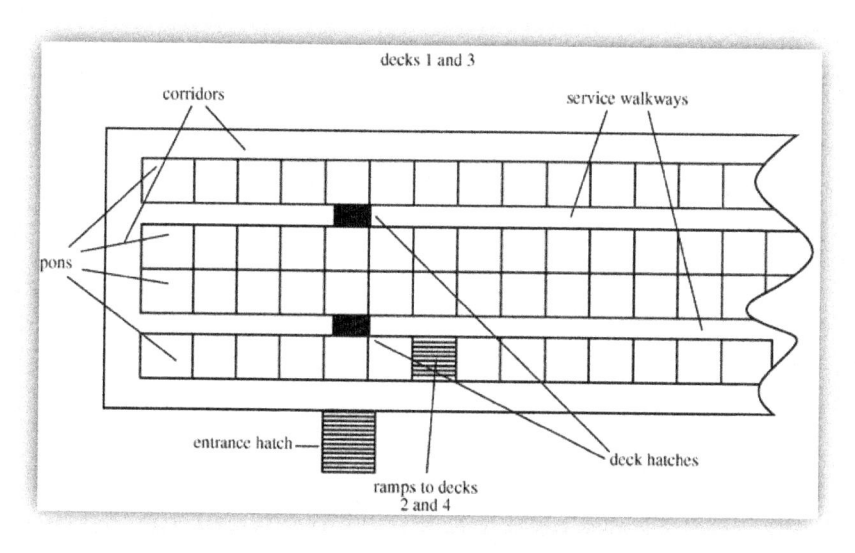

Possible Layout of Noah's Barge—Top View Decks 1 and 3

In terms of construction the wider the hatch was, the more difficult it would have been to close, and therefore, the narrower, the lighter. A larger hatch would have made it more difficult to close tightly to minimize leakage. The hinges along the bottom of the hatch were probably made of several pieces of thick leather from animal hides, but Noah wouldn't have had to worry about his hinges wearing out; he only had to use the hatch twice: once to get on, and once to get off.

If Noah brought one male and one female of each species aboard, any one animal that died before reproducing, would cause extinction for the species. If ninety percent of the animals aboard the barge did not survive (which I will explain further along in the story) then ninety percent of all Earth's living population would have become extinct. Furthermore, one male and one female of a species are doomed to extinction anyhow. The number is too small to allow for gene diversification which encourages healthy offspring.

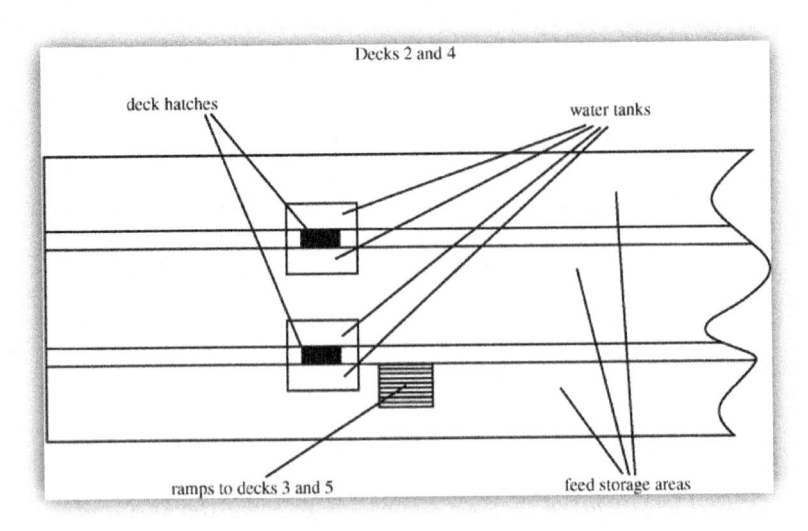

Possible Layout of Noah's Barge—Decks 2 and 4

To take this concept one step further, if Noah and his family were the only people aboard the barge, and the only survivors of the flood, how was the human race perpetuated over the next few thousand years? Did Noah's children, grandchildren and the next hundred generations all marry each other and interbreed? I don't think so, or rather, I hope not.

The explanation has to be that Noah had many other people aboard his barge. These would have been the residents of the village where he lived. Noah certainly needed the help of dozens of people to help him build his barge and then load it with enough food and water required by his passengers, both human and non-human, to last about two months. Noah and his village must have disassembled their boats, houses, barns and fences to enable Noah to build his barge. This would have been a great source of pre-cut and trimmed timber for the barge. The ready supply of building materials would have saved a great deal of time since Noah did not have to cut down trees and haul them by oxen or boat to the barge construction site and then trim them to the desired shape and stack the lumber so it could season.

If the fellow villagers did in fact contribute to the building of the barge, one would expect that Noah would welcome his neighbors aboard his barge so that they too would have a chance of surviving the impending flood. These people would have enabled Noah to maintain his lineage for generations to follow. Indeed, the Holy Bible enumerates the names of a huge number of Noah's descendants.

The people aboard Noah's barge were the ancestors of many people who later lived in Noah's part of the world 7,500 YBP (Years Before Present) and who were relatively unaffected by the Gulf Event. I'm referring to people in North and South America, Northern Europe, Central and Southern Africa, Eastern and Southeastern Asia, and Australia.

During World War II, there were about twenty whooping cranes left on Earth. Years of careful husbandry—and using all the tools of science available at that time—resulted in almost six hundred cranes in 2011. Similarly, when the number of California Condors fell to less than a dozen, the remaining wild ones were captured, and a careful, closely monitored breeding program began. The population grew large enough to release some of the birds into their historic areas. So far the program has been successful.

By 2005, the Ivory-billed Woodpecker population in the United States was apparently gone; the last-seen birds were in eastern Cuba. Around 1990, Ivory-billed Woodpeckers were estimated to be about 100 or more left, but by 2005 none could be found and the last of the Cuban population had disappeared as well.

The Florida panther found mostly in the Everglades National Park has, like many other animals, been teetering on the verge of extinction. An estimated population of up to 200 still exists but inbreeding and a steady mortality rate may be dooming the panthers. The main cause of death for panthers seems to be death by auto collision. Loss of habitat is suspected to be another reason for the decline.

If modern science can achieve such limited successes with preserving endangered species of birds and animals, certainly Noah, if he was able to save only one male and one female of each species, was doomed to failure. But since that did not happen, we can conclude that whatever creatures he had in the barge were there in relatively large numbers.

So if Noah did not have one male and one female of each

species aboard his barge, does that make this part of the story of Noah a falsehood?

No, it doesn't, and here's why. First, think back about the first four books of the New Testament in the Bible. The stories written by Matthew, Mark, Luke, and John all describe the birth of Jesus, based on first-hand observations. Yet, they do not all read exactly alike. These stories have been translated from one language to another, several times over and over again during the last 2,000 years. It is highly likely that something here and there was lost in the translation along the way, either as omissions, different interpretations of the word's meaning and intent, a lack of appropriate available words in the new tongue, or just plain mistakes.

The story of Noah is 7,500 years old, and who knows how many times and into how many languages the story was translated and rewritten over the years.

The phrase "all the animals in the world," may have originally read "all the animals in NOAH's world," which is a perfectly plausible explanation for the story we have today.

The animals he put aboard his barge were probably all farm animals and household pets. Farm animals were highly valued, prized possessions. They were the means to wealth and an indicator of prosperity. This would include dogs, cats, caged birds, pigs, sheep, goats, barnyard fowl, ducks, cows, cattle, maybe horses, and probably oxen to haul timbers around, plus whatever else Noah and his family raised. Farming must have played an important part of their lives, which made farm animals prized and valuable property and farming may have been their primary occupation.

With all these creatures aboard the barge, Noah would have needed a huge amount of water and food for the animals, as well as for the people aboard the barge. Noah probably lived on land bordering a system of lakes and rivers to be able to transport animals and get food to markets. Therefore he, or some members of his family, must have been shipwrights. They probably made small sailing vessels big enough to handle the animals and farm products. The knowledge of how to build a boat could have been the basis for how to construct a barge. So Noah and his siblings, cousins, and children knew about farming and boat building.

Using the Bible's description of the size of the barge that God told Noah to build, it would be a rectangle about 450 feet long, 75 feet wide, and 45 feet high. Its draft is unknown and really not relevant here. The vessel also had about five decks above the bilge with literally hundreds of stalls of various sizes on each deck for the animals. The lowest level of the barge was the bilge, probably a large single room. This bilge is where animal waste, dead animals, and leaking salt water ended up. Noah wouldn't have had any bilge pumps.

The next level up was the first deck, which was really the lowest usable deck. For stability reasons and to keep the center of gravity low, this deck would logically have housed all the large animals. The hatchway would have opened onto this level, so the floor would have been located above the vessel's estimated waterline to try and minimize the amount of sea water that could leak in around the edges of the hatch. Once inside, the smaller animals could have been led up an interior narrower ramp to the third deck where they stayed in smaller stalls.

The second deck would have been designed to hold feed and water for the large animals on the first deck. The stalls on that first deck were probably laid out in long rows running fore and aft inside the hull, allowing room at the bow and stern to lead animals from port to starboard and vice versa as necessary.

The individual, who about a hundred years ago, allegedly saw the inside of the barge where it rests on Mt. Ararat in Turkey, did not give any detailed description of the interior layout that I am aware of. He did, however, make note of many decks and compartments.

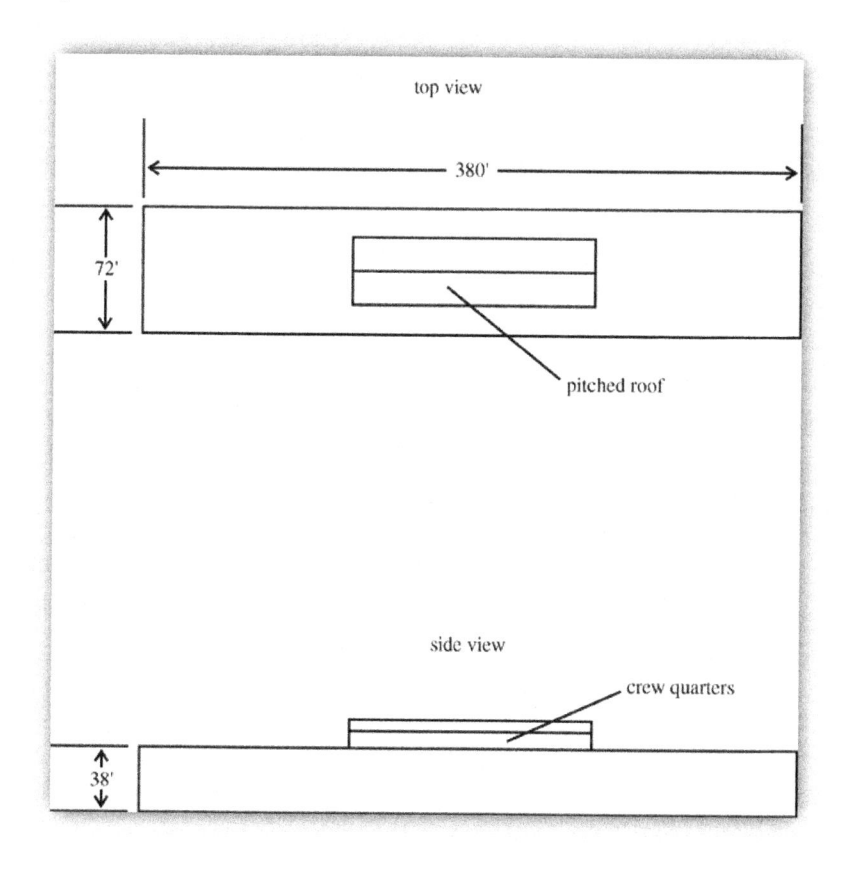

Possible layout of Noah's Barge—Top and side views

I am not a marine architect, but using some common sense, and knowing what the design of the barge was intended to accomplish, here is what I think may have been close to the vessel's layout.

The exterior of the hull was heavy planking of some sort, maybe what was then called wormwood or gopher wood. No one seems to know just what type of wood this was. There is no consensus of opinion, but it must have been some kind of hard, abundant wood, such as oak. The planking was attached to the keel, strakes and ribs, which were made of the same type of wood.

There was no interior planking attached to the ribs because it would not have served any useful purpose, such as to strengthen the hull. A walkway about ten feet wide ran along the outer edge of the deck where it met the ribs of the hull. This walkway would have extended completely around the interior of the barge at this level, with trap doors every thirty feet or so, into which the crew could shovel manure, as well as dead, dying, sick and crippled animals. Torches would be fixed to the ribs here and there to supply light for the animals as well as the crew who worked on that deck.

The stalls may have been large enough to hold several animals each in order to speed up construction time and use less wood.

The second deck would have held the water and feed for the large animals below. There would have been hatches in the floor spaced above the central "feeding walkways" so that food could be shoveled through to the first deck to be distributed daily to the animals and containers of water could be lowered by ropes to the deck below to be poured into troughs between every two stalls. There were also trap doors on the floor of this deck, directly above the doors along the perimeter of the first deck.

The third deck would have been similar to the first but with smaller cages and pens to hold smaller animals. The trap doors in this deck would have been located directly above those on the perimeter of the first and second decks so that manure and dead and dying animals could be shoveled straight down to the bilge, assuming of course the lower deck's trap doors were first opened.

The fourth deck would have held the food and water for the animals and barnyard fowl on the third deck immediately below with appropriate deck hatches similar to those on the second deck for getting food and water to the animals.

The fifth deck would have been the living quarters for Noah, his wife, siblings, parents, children, cousins, and the fellow villagers. This may have been a smaller, raised deck, not covering the area of the entire deck below. The number of people aboard would not have required an area 450 feet by seventy-five feet. Their food and water as well as whatever tools, valuables, and personal property they chose to bring aboard would all have been stowed on this fifth deck.

The smaller animals on the third deck would have suffered similarly to the large animals but on a smaller scale due to their smaller size. To all this, we must add the fact that animals are not good sailors. They are designed to walk and live on solid ground, and even those who wallow or walk into shallow ponds and rivers in the wild still are standing on the ground below. Noah's animals had no prior experience trying to live in close quarters aboard a dark, wildly topsy-turvy, stinking boat while listening to a screaming super hurricane outside.

The scene on the first deck must have been pure mayhem. Animals are not born with an affinity for life at sea in a closed up boat of any size. One of the reasons I think there was around ten feet of headroom on the first deck was so that a panicked, large animal, rearing upon its hind legs, would not have split its head open on the beams supporting the second deck.

The barge was not built with a pointed bow or stern as we might picture it. The only vessels with pointed bows are those built that way to slip through the water easily and thus achieve faster speeds. Noah's barge had no rudder either, because a rudder can only be of use if the vessel is moving through the water faster than at-drift speed. Moving water pushing against a rudder's flank will push the stern to port or starboard, and can only work in a powered vessel, meaning powered by sail, oars, an engine, or a boat being towed.

The barge was designed simply to float at the mercy of the sea and move in whatever direction the wind, waves, and current pushed it. Any vessel drifting on a sea surface which is predominantly either large swells or waves coming from one direction, and not an angry, unsettled sea with winds of twenty knots, (depending on the size of the vessel), will assume a position parallel to the waves. This means that side-to-side rolling would have been the main problem to be reckoned with. Such rolling would have tipped the boat to maybe 60° or so. If the vessel had reached 90° it wouldn't have been able to right itself, and it would then founder.

All five decks would have been equipped with torches for illumination, just as mentioned above for the first deck.

The fifth deck was probably the only deck where hatch

Revisiting Armageddon: Asteroids in the Gulf of Mexico

openings to the exterior were located. This would have given the humans aboard a place to view the sea and sky conditions and would also be the sole source of ventilation in the entire barge. When waves were breaking over the hull any and all hatches would have been closed and secured to avoid taking on any water. I doubt if any vent openings were built-in below this deck for fear that the seas would flood the barge and sink it.

The people aboard the barge would not have had any life jackets or lifeboats because if a barge this size could not weather the storm, certainly a small lifeboat was also doomed. There were no other boats anywhere that could rescue them and no land to try and reach, no sails or engine to propel the barge, no radio, and no Coast Guard. Either the barge would survive the flood, or everything and everyone would be lost.

The fact that the barge did eventually survive, after a roller coaster ride of almost a month and a half, is incredible—almost unbelievable. The design of the hull played no small part in the barge's seaworthiness. A hull with a length to beam ratio of six to one (450 feet to seventy-five feet) or larger is similar to what marine architects use today on many vessels. If well-constructed, such vessels could ride out swells and waves that raised the midsection of the hull leaving the bow and stern out of the water or in very little water, without splitting the hull amidships. Similarly, if the bow was on top of one swell and the stern on the next swell, with no water or very little amidships, the hull would not buckle in the middle and break apart. Noah must have known about these design hazards and how to avoid them by using a proper design.

It is interesting to read about other barges that were constructed about the same time by other people apparently south of Turkey in the Iraq/Iran areas. One such design called for a nearly square hull nearly the size of a football field. The first major swells and huge waves it encountered would have resulted in a complete collapse and breaking up of such a hull.

In the 1800s George Smith of the British Museum took on the job of piecing together and deciphering some clay cuneiform tablets which he found in Nineveh. Some years later he had succeeded, and the story he deciphered was "The Epic of Gilgamesh." The above description of a wide barge appears in this epic, where references are also made about a number of vessels built to try and survive the approaching flood. History does not tell us if any were successful, so we can assume they probably failed.

I find it very difficult to imagine the living conditions for both animals and people aboard Noah's barge. Torches would have provided little light. The screaming of the wind, pounding of the waves, and rolling and pitching of the barge for a month and a half would have thrown the animals into absolute panic. Many would have been thrown down flat in their stalls due to heart attacks, violent seasickness, and especially broken legs and other bones. The other animals in their stalls, which lasted the longest, would have trampled the sick and injured ones to death. Most of the casualties would then have ended up being shoveled or pushed into the bilge by the crew.

A boat with a sea anchor would drift with its bow into the wind and seas. I'm sure, however, that Noah was not aware of sea anchors, or could not have made one big enough to float just under the surface and minimize the rolling of his barge. The barge was probably too long and heavy to be subject to pitch-poling (cart wheeling end over end). However, if the seas were monstrous and the wind blowing at 150 to 300 knots or higher, then all bets are off. The barge would have been tossed around violently in all directions. How any living creature could have survived is a miracle.

I believe that Noah would have figured out the necessity of placing baffles in the bilge. Baffles are solid walls or partitions that extend from port to starboard across the barge. He probably would have constructed two of them to divide the bilge into three compartments. The baffles would keep the weight of the bilge contents more evenly distributed, and maintain the stability of the hull.

Without them, if the seas lifted one end of the barge higher than the other and remained in this position for ten to fifteen seconds, all the sea water that leaked into the bilge, plus the gurry, manure, and dead animals would have been sent flying to the other end by the force of gravity. If the barge held two to six feet of sea water, the force of it would have burst open the lower end of the hull and sank it. Even if the hull stayed intact, the weight of all this material in one end would have stood the barge almost on end, which also would have caused it to go under, end first and very rapidly.

There is still another factor to be pondered by us, and that is the terrible stench that must have permeated every deck, and therefore, contributed to further illness and death. The lack of ventilation would have trapped all the smells tightly in place throughout the barge.

By the time the wind abated and the seas flattened out and the barge grounded on Mt. Ararat, the animal mortality rate must have been severe. I would guess at ninety percent or more among the large animals and less among the other animals. The smaller animals on the third deck would have had a higher survival rate due to their size.

Noah must have anticipated so many deaths and maybe this is one reason why he built the barge so big, so that even a small survival rate would result in enough animals to breed and allow their species to multiply and survive.

Because we now know that the direction of the wind and flood waters was easterly, it gives us a clue as to where Noah lived, and where he built his barge. If he had lived near Mt. Ararat, or anywhere nearby such as another mountain in Turkey or Armenia, he would have ended up somewhere between the Caspian Sea and the western end of the Gobi Desert in Mongolia. The easterly flowing stream, in forty days, would have taken him far to the east. The storm certainly did not push him in circles around eastern Turkey, and then plant him on Mt. Ararat.

Noah must have come from a place far to the west of Ararat. If the storm was as wild as we think, the barge would have drifted hundreds of miles in forty days. It is impossible to know or calculate his daily drift. If the storm pushed him about fifty miles per day, this means he came from a place about 2,000 miles west of Ararat.

The ferocity of the storm would have been at its worst around the Gibraltar/Atlas Mountains area near where it first made landfall. The forces of gravity and obstructions of mountains and hills slowly reduced the easterly speed of the storm waves as well as the wind, which leads me to believe that Noah could not have lived in the western Mediterranean area, where the storm's ferocity peaked. The chances his barge could survive here may have been less than if he lived in the eastern Mediterranean area somewhere. To speculate a bit, a starting point about 2,000 miles to the west of

Revisiting Armageddon: Asteroids in the Gulf of Mexico

Ararat would have placed Noah's home somewhere near Malta or Italy.

One thing that has puzzled me is the timeframe involving the barge construction. The Bible says that God warned Noah of the impending flood and told him to build a barge in which he could survive the flood. But the length of time between the warning and the flood is unknown. Certainly Noah could not build it if the front wall of the flood was rapidly approaching. He had to have at least a year or longer to construct the barge, find and assemble the animals, get them aboard with the necessary food and water including the needs of the people aboard. I can find no source material that addresses this question.

There may be a more mundane explanation, and the more I think about it, the more plausible it becomes.

We know that the Gulf Asteroid was heading ENE: the Mediterranean Basin or Plains, was in its path. Months or years before the asteroid landed, I think it was in a low Earth orbit—going ENE over the same parts of Earth on a regular basis. For purposes of simplification, let us assume it orbited once daily. Noah and his people would have noticed on a clear day or night a small speck coming from the western horizon, going overhead and disappearing at the eastern horizon. How did it get back to the western horizon each day? Did it stop and reverse course and then proceed easterly again?

I think Noah was smart enough to figure out that Earth, like the moon and the sun, was round, and the speck was orbiting Earth. This is the only explanation that makes any sense.

As time went by, the speck grew steadily larger, until it may have been close enough for Noah to detect other small asteroids that went along with the big one.

Eventually, the speck grew so big as it got closer to Earth that it may have blocked out the sun for a few moments. Its orbit was slowly decaying as it got closer and closer. About this time God spoke to Noah about the oncoming flood and instructed him to build an ark. Noah and his people must have been frightened and panicked to see such an object approaching them.

The next thing that crosses my mind is why did Noah build a barge and not a fortress or mountain cave? He apparently was convinced that the problem would come in the form of a flood, as God predicted, and not an ordinary land-only earthquake. How did he know this?

If we put aside the instructions to build a barge that came from God, then here is an explanation that is plausible to me: it could mesh with the word coming from God, because I think Noah knew ahead of time that a disaster was looming in his near future.

Plato's "Critias" goes into great detail about the lives and accomplishments of the people of Atlantis. Plato relates that they engaged in trade with countries within the isthmus of Gibraltar as well as outside countries. He says that just inside the isthmus there was a big lake, which drained into the Atlantic via a small river. Such a lake would probably have a number of rivers feeding into it from sources in the Mediterranean Plains region coming south and west from France and Italy, as well as from the Sahara river system, and maybe rivers coming from the area around Malta.

The Nile and other rivers leading into the eastern Mediterranean probably led to the Arabian Sea via the Red Sea. These river systems must have been navigable by the biremes, triremes and other sailing vessels of that era. This means, that

Revisiting Armageddon: Asteroids in the Gulf of Mexico

Noah could have sold cattle and other animals and farm products to buyers in the Western Mediterranean area. Certainly, he would have known about Atlantis. In any case, the word would have reached Noah that the Atlanteans knew about the great ocean that lay to their west.

Noah could thus conclude that the impending crash of the asteroid he was looking at every day could cause a big flood if it landed in the ocean, still heading in its easterly direction toward the low flat plains of the Mediterranean area, and consequently, Noah's home.

For these reasons, I think Noah built a barge and not a stone fort. When God warned Noah to get going on the construction of a barge of some sort, His word would act as reinforcement of Noah's own fear of an impending flood.

And now you know when, where, how and why the barge was built.

Chapter Thirty-Four

Atlantis

At this point in the story, you may have a pretty good idea where Atlantis was located. Over hundreds of years, close to 5,000 books and articles have been written about the mysterious, almost mythical island-continent called Atlantis, which allegedly "disappeared under the sea" long ago in some sort of calamitous natural disaster which befell that area.

One of the earliest people to talk about Atlantis was Plato, a philosopher of ancient Greece. Up until now, the exact location of Atlantis has been a matter of wild conjecture.

The general thinking has Atlantis centered at a location somewhere in the Atlantic Ocean. Some reports place it in the eastern North Atlantic; others have it in the Bahaman Islands, and still others at a place in the western Atlantic Ocean, the Pacific Ocean, and even in the Andes Mountains of South America.

Plato wrote two stories or "dialogues," as he called them, in about 400 BC His method of writing was to use a fictional conversation between two people, somewhat like a playwright would use. They were called the "Dialogues of Plato."

Both dialogues involved Socrates, mentor of Plato, as one of the characters. The first dialogue's other person was Timaeus, but only the end of this dialogue made any reference to Atlantis.

The second dialogue involved Socrates and Critias, a relation of Plato. Critias said he learned of Atlantis from his grandfather, who heard it from Solon, who heard it from Egyptian priests he met during his travels to Egypt. Solon was one of the wisest men of ancient Greece.

This second dialogue covered most of the story of Atlantis. In the dialogue, Solon asks the Egyptian priests what they knew about antiquity because he knew very little about the old times. The priests indicate that they had in their possession ancient records of both the city/state of Athens and Atlantis.

One of the priests then tells Solon all he knows. Solon then tells the story to Critias, who passes it along to Socrates, and then to Plato. In Timaeus, the priest states that Athens was a great and prosperous city 9,000 years ago and that the ancient Egyptian city of Sais was 1,000 years younger.

At the end of these dialogues, a war is alluded to, which Critias says was the time of Athens and other powers east of Gibraltar defeating Atlantis, and that this war took place 9,000 years ago. (That would be 11,400 YBP.)

The only explanation of this apparent discrepancy would be that Athens began and matured within 1,000 years and then the war began which makes the founding of Athens some time before 11,400 YBP. This was all recorded in the Egyptian "sacred registers."

The priest then tells about a mighty power which, without any provocation, invaded Europe and Asia, but with the help of the Egyptians was defeated by the Athenians. The invaders came from an island nation called Atlantis, situated outside of the "pillars of Heracles." This refers to the Gibraltar area.

Further on, the dialogue states that Atlantis was "the way to other islands" which could be a reference to the Caribbean, and from there one can go to "the whole of the opposite continent which surrounded the true ocean." This opposite continent could indeed be North America. There was a "sea within the straits of Heracles" (meaning to the east of Gibraltar), but "which had a harbor with a narrow entrance."

The character of Critias then gives us a small clue as to the size of Atlantis. He states that the outer perimeter of the central plain that included the city was a rectangle 3,000 by 2,000 stades, or a circumference of 10,000 stades. This is the only reference he makes to a measure of length or area.

Dictionaries state that the accepted length of a stade is about 607 feet. If so, then 10,000 stades = 6,079,000 feet. This equates to about 1,150 miles in perimeter, which is equivalent to a square 325 miles on each side. Let's round the area off at 105,000 square miles, which would be a very rough estimate. However, if the aspect ratio of this central plains area was 3:2, then the total area would be reduced to about 78,000 square miles, or slightly smaller than the area of the State of Colorado.

If the ten kingdoms were equal in size, then each kingdom would be around 10,500 square miles in area in the first instance above, or 7,800 miles in the second more rectangular size. Add to this the remainder of the island outside the central plains and we end up with a much larger island.

In "Critias," Plato makes mention of mountains and rivers, only he does not give us a clue as to the area of this remainder portion. Mountains and rivers do not form in small areas, so I think the total area must have been at least double or triple the area of the central plain. Plato makes mention of the land trailing off in a southwest direction.

The Atlanteans invaded Athens, (Greece), Egypt, and "the entire region within the straits," (Spain, Portugal, Basqueland, Morocco, Tunisia, Libya, and Italy). Apparently, this did not include Turkey and other countries or areas in the Middle East, as we know it today, nor those in the Balkans, except Greece.

He then tells that later "there occurred violent earthquakes and floods, and in one day and night the warriors and the island of Atlantis disappeared into the sea, leaving the sea impassable due to a shoal of mud in the way, which was caused by the subsidence of the island."

I believe that the shoal was created by the severe washing away of Atlantis. The fact that today, 7,500 years later, this area is now navigable by ships, is explained by the southerly currents flowing along the Atlantic Ocean's east side, which would have slowly taken up this mud, carried it along, and then dumped it in the eastern area of the South Atlantic Ocean, along the coast of Africa.

I feel that the "subsidence of Atlantis" elsewhere referred to as the island sinking beneath the sea, is not exactly correct. There was nothing gentle about it.

The Gulf floodwaters racing across the Atlantic struck Atlantis head on and literally tore it apart and washed it into the Mediterranean Basin, depositing parts of it everywhere the water slowed down, from the area just west of Gibraltar to the central portions of the Mediterranean Sea. Much of the debris was dumped into the Adriatic Sea, making it much shallower than the seas west of Italy.

This is why nothing of Atlantis exists today anywhere. The force of the flood pulverized just about everything even including some bedrock. Anything alive, meaning birds, animals and people were crushed or drowned and things made of wood and stone likewise were destroyed. Other things just rotted away. There could be artifacts such as metal manmade objects or building stones buried in the Mediterranean region, or more likely deep under the bottom of the Adriatic Sea. Here, they could be thousands of feet under the sea bed where it is highly unlikely they will ever be found.

If any man-made objects did survive, the main reason we never found them is because we have looked in the wrong places. I think any such objects, if in fact they do exist, would be found buried under the seabed west of Gibraltar, the seabed east of Tunisia, and more likely buried, perhaps thousands of feet down under the floor of the Adriatic Sea, or under the sea beds of the Black and Caspian Seas.

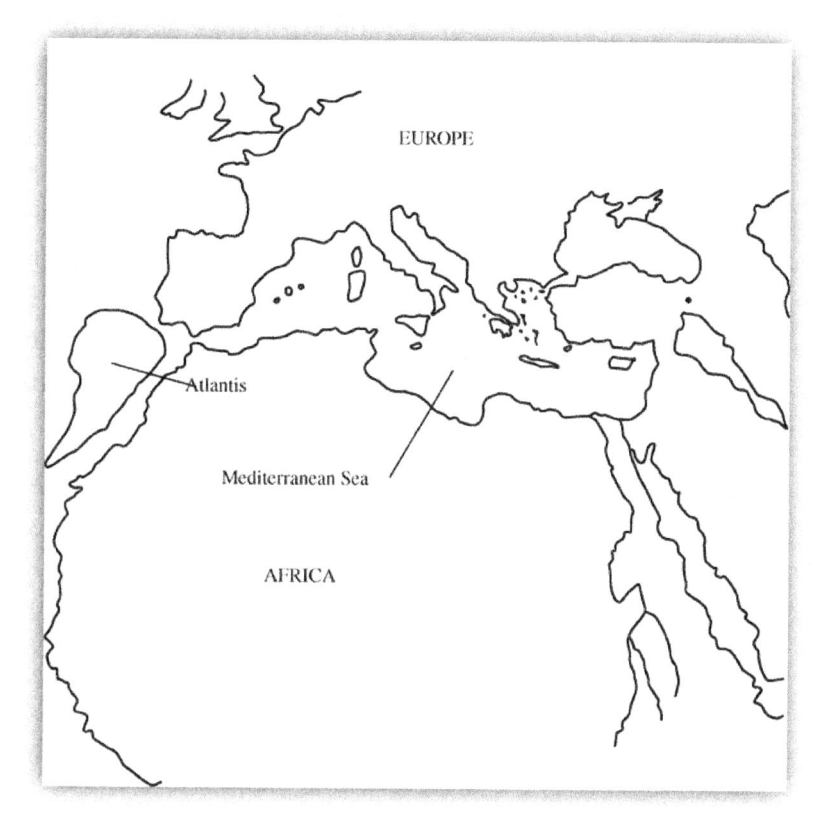

Revisiting Armageddon: Asteroids in the Gulf of Mexico

Possible location and size of Atlantis prior to its destruction 7,500 YBP. Neighboring lands shown as they appear today to aid identification

I believe there is one paramount reason why many people consider the entire Atlantis concept to be a myth, and that is simply the fact that tangible evidence—either artifacts or the land itself—no longer exists.

As I mentioned earlier, the Gulf Event, arriving at Atlantis, was not just one big wave. It was, indeed, one wave, but was several thousand miles long, from front to rear. For all this water to come through the Mediterranean area must have taken several days, if not weeks. The Canary, Cape Verde, and Azores Islands must have been much higher in elevation than they are today. The lower areas were also eroded, and ended up underwater when it was all over. The island "mountains" were cut way down in size.

Certainly, the Atlas Mountains, which stood right in the central path of the flood, were badly eroded and reduced in size. That may be very obvious when we consider that these mountains were one of the primary sources of the headwaters of some of the river systems that flowed through North Africa and into the Red Sea. Other western rivers may have flowed through the tiny strait leading to the North Atlantic Ocean west of Gibraltar.

The enormous amount of sand, mud and silt the flood scoured out ended up partially to the northeast of Tunisia, and more so into the Adriatic Sea, Aegean Sea, Black Sea, Caspian Sea, and into Ukraine as well, creating the breadbasket of Russia in that area.

If the Great Flood had simply just washed over the land, and not scoured it away, such as in Spain, Italy and Libya, no myth would have ever developed.

Atlantis stood directly in front of the advancing wall of water I call the Gulf Event Flood, and the island continent was ripped up and washed into the Mediterranean area along with the entire isthmus of Gibraltar. There are no tangible remains left to hold, look at, or ponder over.

But yet, many believe Atlantis did exist, as Plato relates, and many believe it is all a myth created by Plato's imagination. The choice is yours to make.

CHAPTER THIRTY-FIVE

The Thief That Stole Mars

If you have comprehended everything you have read so far, congratulations, and cheer up—this is the next to last chapter.

Your head is probably burning by now with a few questions that you would like to ask me. I think I know what they are, and so I'll try to read your mind and answer them. If I miss any, sorry.

To do this the simplest way will probably be an ordinary question and answer process, so here we go:

Q. Are you for real? Are the things I have read all true?

A. Yes. After thinking about the subjects I discussed, I tried to analyze all the possible solutions I could think of to the mysteries involved. There are two matters that border on fiction, but my purpose for them is solely to define and try to explain a couple of things that you may have difficulty comprehending.

One is the chapter I call "A Great Ship Goes Missing" and the following six chapters, and the other is the details I describe about the probable construction of Noah's barge. I believe everything else is the truth, but I must admit it is sometimes difficult to prove what I say.

Looking at any one piece of evidence about solving a mystery is meaningless, but when you read about many of the pieces of the puzzle that fit, the truth surges forward.

Q. How long did it take you to research and write this book?

A. Nine years. Most of what I wrote is all original thinking. Based on what I could find out about these mysteries, and using a little common sense, the answers in many cases just jumped out at me. Quite often the solutions presented more questions, and in some cases they answered things I never thought about, such as where did Noah come from, or why are there hammocks and cracks in the limestone cover in certain places in the Everglades.

But I didn't come anywhere near spending nine years full time on this project. I still have a business to run, a family to enjoy, civic things I got involved in, and some recreational activities to pursue.

Most of the time I spent writing, rewriting, editing, and revising over and over again was spent in the cabin of my boat on summer days in Massachusetts, or late at night in the lanai of my winter home in Florida.

I need two or three hours at a time with no interruptions to work on this story—to be alone with no TV, no telephone, no knocks on the door, and no outside noises to speak of.

After two or three hours, my head gets worn out and tells my body to knock off and get some sleep, and so I do.

Revisiting Armageddon: Asteroids in the Gulf of Mexico

Q. What qualifications do you have to write a book like this?

A. None at all. Maybe I should amend that. I had a great English teacher in secondary school, but I did not take English courses in college because they offered none. I suppose the university reasoned that if I had not figured out the English language by the time I completed secondary school then I don't belong in college, and I'd agree with them.

My dad taught me his style of letter writing, which was very informal. I try to make my writing read as though you and I were alone together and I was telling you this story orally. And to me, the simpler the better, with no irrelevant garbage. I find that magazines and newspaper articles often start out to be very boring. I usually skip or glance over the first two or three paragraphs and try to start with what the author is really talking about. Those first few paragraphs are often just pointless, irrelevant baloney.

Q. What other books or articles have you written?

A. Again, none. What you just read about in this book was burning a hole in my mind. It had to be written down.

Q. Do you plan to write any other books?

A. Maybe, yes. One would be another geology type book, but quite different than this one. It goes back to an occurrence that took place maybe 100,000 to 500,000 years ago; I haven't figured it out yet. Even more so than in this book, source material is completely absent.

The other book would be a novel on a subject matter that, as far as I know, no one has ever written about. But that's down the road a few years, if ever.

Q. Could a Gulf type event occur again? If so, when and where?

A. Yes, there could well be another Gulf event that hits planet Earth, but probably at a different location, although I'm not sure where.

As to when, that I don't know either. We will probably be hit by something so big we can't avoid it, perhaps in the next 100,000 years or so, more or less.

There has been a lot of talk and predictions by the ancient Mayans, Edgar Cayce, Nostradamus, and others, to the effect that an asteroid would strike Earth in about December 2012 or later, causing disastrous results to our civilization.

That date came and passed uneventfully. But if these predictions, made long ago, were off by a few years, perhaps something could happen.

As time trickles along its slow path for months and years, this entire matter tends, like any bad thought, to park itself in the dead storage warehouse area in the back of our minds, and brought to the front less and less frequently as time presses until it is nearly forgotten for good at some time in the future.

Cayce inferred that there was no certainty that we would be doomed, because we could try to cause the asteroid to avoid us and maybe nothing would happen. That is, assuming we could find it, track its path, and take measures to either obliterate it with atomic explosives, or push it off its course enough to miss us.

Neither of these responses is a guaranteed solution to the problem. If we could find such an asteroid, and it turns out to be small enough, maybe ten or so miles in diameter, and we peppered

Revisiting Armageddon: Asteroids in the Gulf of Mexico

it with atomic explosions, Earth would then be subject to a bombardment of small asteroids which could be disastrous.

On the other hand, if we could instigate a cluster bomb mentality with explosions not close enough to disintegrate it, but rather to push it off course a bit so that it misses planet Earth, this process could possibly be the better of the two solutions. It all depends on whether or not we find it in time and can carry out the effort. A group assault by United States and Russian missiles, working together, similar to our space station cooperative efforts, may be necessary.

Pushing it off course by a tiny amount, if it occurs millions of miles out, could result in an off course track of many miles by the time it arrives near us.

However, if it should creep up on us closely and pass Earth by a small margin, the result could wipe out all life on Earth. I don't know how close a path is safe for us. It could be 100 miles, or maybe 10,000 miles would be a minimum safe distance. It would take some brilliant work by astronomers and physicists, far more knowledgeable than me, to come up with a workable plan.

The variables involved are the asteroid's size, mass, composition, path, time frame, and speed relative to Earth. If luck fails us in every one of these factors, then we are doomed. Personally, I think that is very unlikely to be the case.

Q. How do I know that under such a set of unfavorable conditions life of Earth could end?

A. The answer to the question is that this type of disaster has already occurred in our planetary system. I'm referring to Mars.

Ray Covill

Space vehicles that NASA has sent to Mars have returned some amazing pictures of the surface. It appears to be covered with various size stones and rocks. The interesting thing is that Mars is not pockmarked with impact craters of all sizes such as on our moon. Neither is planet Earth, in this case because our atmosphere has burned up all the smaller ones, but our moon never had an atmosphere to protect its surface.

Scientists say our solar system is about 4.6 billion years old. It is, therefore, reasonably safe to say that the age of Mars, Earth, and our moon are all about the same.

The assemblage of space junk that developed into planet Earth was actually a tiny bit larger than Earth. A small amount was located sufficiently apart from the main group of materials that formed Earth, to not become part of Earth. As the majority of rocks tumbled together due to the forces of gravity, it formed a 7,000 mile diameter sphere.

The remainder was distant enough to form its own sphere, (our moon), which is about 1% the size of Earth, but too small and light to develop a molten interior, or develop an atmosphere or an ocean.

I would expect that most of our moon's impact craters date back to its first million years, or however long it took for most of the debris in space to find a home on all the planets and moons and leave our interplanetary space relatively clean.

If Mars had no atmosphere since its formation, it should appear just as pock-marked as our moon. But that is not the case. The absence of such a surface means that Mars once had an atmosphere, but lost it. Also, if Earth developed an atmosphere, Mars most likely did, also. Both planets are of similar size and

Revisiting Armageddon: Asteroids in the Gulf of Mexico

distance from our sun.

If Earth also developed an ocean, I think Mars probably did too. We have just about as much proof of this as we do about a Martian atmosphere.

Here's what I think happened, quite suddenly, maybe two or three billion years ago, when very early life forms first began to appear in Martian seas, the same as occurred on Earth.

Another assemblage of space junk, similar to what developed into Earth and our moon, appeared in a solar orbit not too far from Earth's orbit, and at about the same time. This one was also split into two groupings similar to the ones that formed Earth and its moon.

This other one was not in a ratio of 99 to 1, but closer to 55/45 %. The main grouping eventually sucked in all nearby space material and formed Mars.

The smaller portion also became a small planet, planet "X", let's call it, moving along in the same exact orbit as Mars, but was located a bit out ahead of Mars because of its dense mass and faster "sling speed." It probably contained a higher iron content than Mars.

It slowly pulled away from Mars, leading the way around the Martian orbit. As millions of Martian years passed, X eventually became situated in an orbit exactly opposite Mars—half way ahead of it. It now ceased to widen the gap, and began to instead slowly close the gap. Eventually X caught up with Mars. It finally got within fifty to 100 miles of Mars, and went into a very slow orbit around Mars. Because of X's density, it had a much stronger gravitational force than Mars. As it approached Mars, a strange thing happened.

The gravity of X began to pull the Martian air, or whatever composition Martian "air" had, up toward X. This occurred because the gravitational pull of X was much greater than that of Mars. The "pulling" on Mars' surface would have resulted in hurricane force winds starting up and then stopping as X revolved around another side of Mars. At no point did it ever collide with Mars.

In addition to air, the lightest part of Mars, the gravity of X pulled away Mars' water—be it seas, lakes, or rivers. Only underground water survived as is found today at Mars' poles.

Water came rushing across the Mars surface in great flood fashion and raced toward the nearest "pickup point," where it literally rained upside down. The rain started on Mars' surface and was pulled up toward X as it reached the closest point of X in each orbit around Mars.

The wild racing of water across Mars' surface created deep gullies, and formed the dry stream beds which we see there today. The rushing water picked up a great deal of dust and dirt en route to its ascending locations, as we might expect.

X did not stop at any one location over Mars, but rather, it circled Mars. So the dirty water, by now frozen into dirty ice, sort of fell in behind X and started to trail X. The air simply disappeared into space.

After a few orbits around Mars, X finally started to accelerate as it grew heavier with the Martian dirty ice it continued to collect on each orbit. Eventually X was whipped away from Mars by the same "sling shot" effect we have learned about when

Revisiting Armageddon: Asteroids in the Gulf of Mexico

an Earth satellite orbits closely around our moon, and is hurtled back toward Earth.

So now we have X heading off into space in a long orbit around the sun, and trailing behind it is a long line of dirty ice crystals, spread over thousands of miles. At some point soon after pulling away from Mars, the slingshot effect of X accelerated it too fast for the trail to keep up with it, and the trail was left behind, still in the same solar orbit as X had been, but slower. X now either dove into our sun or one of the other planets that revolves around the sun, or went into some new orbit as a new planet, or flew out of our solar system completely.

We will never know if it flew into our sun. There is no evidence to show this, except for the large sunspot visible from Earth, whose origin is unknown and conjectural. The chance that it flew into another planet is very small—too small to be of any significant probability. It is also highly unlikely it could have flown out of our solar system into interstellar space simply because I doubt if it could have developed enough slingshot speed to do so, even under the most favorable circumstances.

If it did not fly into the sun, the only other answer would be that it went into a new orbit around the sun. Could it be planet Venus or Mercury? We'll never know, but this explanation along with its flying into the sun, are two plausible explanations.

Meanwhile, the stream of dirty ice crystals continued to orbit the sun, but at a slower rate because of its small mass. That stream of ice crystals is what we call a comet. Because of the nature of its birth, I would guess it is one of the larger comets now in orbit around our sun. It is even possible that it is Halley's Comet. This comet orbits the sun every seventy-five years or so and the latest pass visible from Earth was only a few years ago.

I believe NASA sent a space vehicle into this comet to capture some of the crystals, but I'm not sure if they were analyzed in space by our space vehicle or whether a sample was sent back to Earth.

It would be very interesting if scientists could analyze the structural and chemical composition of dust and ice on Mars and compare it to those of the dirty ice crystals in the comet. If they turned out to be a match, then this idea of Mars losing its air, water and dust, would be proven true.

This is a long answer to a short question.

Q. You have explained a number of mysteries that have puzzled people for a long time. Are there any other ones you could tell us?

A. Not at this time. But the biggest of all and one that I cannot explain is why after 7,500 years since the Gulf Event, has no one else explained the things you have read about in this book. Billions of people have come and gone over this span, but why has no one found the answers until now? There are, and have been, millions of people far more intelligent than me, who have not come up with the answers.

So my big question is, "Why me?" And to that, I have no answer.

CHAPTER THIRTY-SIX

Final Thoughts

The first streaks of dawn are not far away. Just over the horizon. I'm sitting here on my lanai, winding up this story. The night is black. The air is still, broken occasionally by the first, early calls of the collared doves. The damp smell of night is beginning to lift.

Now the sky is getting brighter. Blurred light is sneaking through the holes in the screens. My back is tired and my eyes need propping open.

My arm is reaching deep into my skull to scrape up these last words and then shove them down my hand to my fingers, which scrawl them onto this pad. Now my eyes are burning.

The accelerating brightness is finally giving way to high-level, warming rays of sunshine.

For no apparent reason, I am looking down at my body and I suddenly realize it is now Saturday…

All over my body.
A new day has begun.
And the sun is shining.

The End

Ray Covill

ABOUT RAY COVILL

Ray Covill was born and raised in the Fairhaven/New Bedford area of Massachusetts. Birding and boating have been among his favorite pastimes over the years. His passion for birding continues to lead Ray and his wife on extensive travels throughout both North and Central America.

It was sometime during the 1960s or 1970s when Ray watched a television program that addressed the Bible's claim that The Great Flood covered all of the earth. The program proposed

Revisiting Armageddon: Asteroids in the Gulf of Mexico

that Earth suddenly split open and poured out sea water that covered the world—and then, it tightly closed itself up again. It later opened up once more, and the excess seawater went back underground. This not only seemed absurd, but quite impossible to him.

Ray finally concluded that the original text must have read that the flood covered all of the area in *Noah's* world. Now, there was an answer he could really believe. This realization was the inspiration and the motivation for him to put pen to paper and write his debut work, *Revisiting Armageddon*.